ADVANCED SURVEYING
(THEORY & PRACTICE)

AUTHORS

Dr. Ramakant Agrawal
Professor and Head,
Department of Civil Engineering,
Oriental Institute of Science and Technology, Bhopal

Mr. Parshottam Sarathe
Asstt. Professor,
Department of Civil Engineering,
Oriental Institute of Science and Technology, Bhopal

REVIEWER

Dr. Sirisha Uppaluri
Associate Professor,
Department of Civil Engineering
Christ (Deemed To Be University), Bengaluru

KHANNA BOOK PUBLISHING CO. (P) Ltd.
4C/4344, Ansari Road, Darya Ganj, New Delhi, India

All India Council for Technical Education
Nelson Mandela Marg, Vasant Kunj, New Delhi, India

Dear Readers,

To prevent the piracy, this book is secured with HIGH SECURITY HOLOGRAM on the front title cover. In case you don't find the hologram on the front cover title, please write us to at contact@khannabooks.com or whatsapp us at +91-99109 09320 and avail special gift voucher for yourself.

Specimen of Hologram on front Cover title:

Moreover, there is a SPECIAL DISCOUNT COUPON for you with EVERY HOLOGRAM.

How to avail this SPECIAL DISCOUNT:

Step 1: Scratch the hologram

Step 2: Under the scratch area, your "coupon code" is available

Step 3: Logon to www.khannabooks.com

Step 4: Use your "coupon code" in the shopping cart and get your copy at a special discount

Step 5: Enjoy your reading!

ISBN: 978-93-5538-510-9
Book Code: DIP460EN

Advanced Surveying: Theory & Practice
by Dr. Ramakant Agrawal & Parshottam Sarathe

First Edition: 2026

Originally Published by:
All India Council for Technical Education, Ministry of Education, GOI

Exclusively Published & Distributed by:
Khanna Book Publishing Co. (P) Ltd.
4C/4344, Ansari Road, Daryaganj,
New Delhi (INDIA)

Visit us at: **www.khannabooks.com**
Write us at: **contact@khannabooks.com**

To view complete list of books, Please scan the QR Code:

KPH

Printed in Bharat.

प्रो. म. जगदीश कुमार
अध्यक्ष
Prof. M. Jagadesh Kumar
Chairman

सत्यमेव जयते

75 आज़ादी का अमृत महोत्सव

अखिल भारतीय तकनीकी शिक्षा परिषद्
(भारत सरकार का एक सांविधिक निकाय)
(शिक्षा मंत्रालय, भारत सरकार)
नेल्सन मंडेला मार्ग, वसंत कुंज, नई दिल्ली-110070
दूरभाष : 011-26131498
ई-मेल : chairman@aicte-india.org

ALL INDIA COUNCIL FOR TECHNICAL EDUCATION
(A STATUTORY BODY OF THE GOVT. OF INDIA)
(Ministry of Education, Govt. of India)
Nelson Mandela Marg, Vasant Kunj, New Delhi-110070
Phone : 011-26131498
E-mail : chairman@aicte-india.org

FOREWORD

Engineers are the backbone of the modern society. It is through them that engineering marvels have happened and improved quality of life across the world. They have driven humanity towards greater heights in a more evolved and unprecedented manner.

The All India Council for Technical Education (AICTE), led from the front and assisted students, faculty & institutions in every possible manner towards the strengthening of the technical education in the country. AICTE is always working towards promoting quality Technical Education to make India a modern developed nation with the integration of modern knowledge & traditional knowledge for the welfare of mankind.

An array of initiatives have been taken by AICTE in last decade which have been accelerate now by the National Education Policy (NEP)2020. The implementation of NEP under the visionary leadership of Hon'ble Prime Minister of India envisages the provision for education in regional languages to all, thereby ensuring that every graduate becomes competent enough and is in a position to contribute towards the national growth and development through innovation & entrepreneurship.

One of the spheres where AICTE had been relentlessly working since 2021-22 is providing high quality books prepared and translated by eminent educators in various Indian languages to its engineering students at Under Graduate & Diploma level. For the second year students, AICTE has identified 87 books at Under Graduate and Diploma Level courses, for translation in 12 Indian languages - Hindi, Tamil, Gujarati, Odia, Bengali, Kannada, Urdu, Punjabi, Telugu, Marathi, Assamese & Malayalam. In addition to the English medium, the 1044 books in different Indian Languages are going to support to engineering students to learn in their mother tongue. Currently, there are 39 institutions in 11 states offering courses in Indian languages in 7 disciplines like Biomedical Engineering, Civil Engineering, Computer Science & Engineering, Electrical Engineering, Electronics & Communication Engineering, Information Technology Engineering & Mechanical Engineering, Architecture, and Interior Designing. This will become possible due to active involvement and support of universities/institutions in different states.

On behalf of AICTE, I express sincere gratitude to all distinguished authors, reviewers and translators from different IITs, NITs and other institutions for their admirable contribution in a very short span of time.

AICTE is confident that these out comes based books with their rich content will help technical students master the subjects with factor comprehension and greater ease.

(Prof. M. Jagadesh Kumar)

ACKNOWLEDGEMENT

The authors are grateful to the authorities of AICTE, particularly Prof. Anil D. Sahasrabudhe, Chairman; Prof. M. P. Poonia, Vice-Chairman; Prof. Rajive Kumar, Member-Secretary and Dr Amit Kumar Srivastava, Director, Faculty Development Cell for their planning to publish the books on (Advanced Surveying-Theory&Practice). We sincerely acknowledge the valuable contributions of the reviewer of the book Dr. Sirisha Uppaluri, Associate Professor, Dept. of Civil Engineering, Christ (Deemed To Be University), Bengaluru- 560074, Karnataka for giving valuable inputs and suggestionsto bring this book in present form.

The authors are thankful to Dr. Sadhna Agrawal, Amber, Akshat, Ms.Swati and Aryansh for their cooperation and patience during writing this book.

This book is an outcome of various suggestions of AICTE members, experts and authors who shared their opinion and thought to further develop the engineering education in our country. Acknowledgements are due to the contributors and different workers in this field whose published books, review articles, papers, photographs, footnotes, references and other valuable information enriched us at the time of writing the book.

Dr. Ramakant Agrawal

Parshottam Sarathe

PREFACE

The book titled "Advanced Surveying – Theory & Practice" is an effort of the authors to present their vast experience of teaching in the simplest form specially for diploma students strictly as per syllabus prescribed by AICTE New Delhi. The level and presentation of the book is kept in such a manner that diploma students can understand the subject easily. Efforts are made to explain the fundamentals of the subject in the simplest possible way.

During the writing of various chapters of the book, various standard textbooks, web materials and you tube content were referred and accordingly we have divided the book into six chapters. The subject matter is clearly explained through definitions, principles, illustrations, and figures. The solved and unsolved numerical problems are included to explain the subject clearly, wherever necessary. The multiple-choice questions, short and long questions are included for preparations of various types of the examinations.

Know more section is included in each chapter to teach the students the content beyond the syllabus. This will enable the students to learn latest knowledge, supplementary knowledge and history to update themselves and making the content interesting. Dynamic QR codes are provided at the end of each chapter to learn the subject in depth including interesting Videos to explain the subject elaborately and clearly.

We expect that the book will be useful for diploma students and will inspirethem to learn and discuss the applications of various surveying instruments and techniques. Every effort is made to minimise the errors in the content and numerical work, in spite that some errors may remain, and authors would be thankful to the readers if they point out the same.We would appreciate allconstructive comments and suggestions which will contribute to the improvement of the book

Dr. Ramakant Agrawal

Parshottam Sarathe

OUTCOME BASED EDUCATION

For the implementation of an outcome based education the first requirement is to develop an outcome based curriculum and incorporate an outcome based assessment in the education system. By going through outcome based assessments, evaluators will be able to evaluate whether the students have achieved the outlined standard, specific and measurable outcomes. With the proper incorporation of outcome based education there will be a definite commitment to achieve a minimum standard for all learners without giving up at any level. At the end of the programme running with the aid of outcome based education, a student will be able to arrive at the following outcomes:

Programme Outcomes (POs) are statements that describe what students are expected to know and be able to do upon graduating from the program. These relate to the skills, knowledge, analytical ability attitude and behaviour that students acquire through the program. The POs essentially indicate what the students can do from subject-wise knowledge acquired by them during the program. As such, POs define the professional profile of an engineering diploma graduate.

National Board of Accreditation (NBA) has defined the following seven POs for an Engineering diploma graduate:

PO1. Basic and Discipline specific knowledge: Apply knowledge of basic mathematics, science and engineering fundamentals and engineering specialization to solve the engineering problems.

PO2. Problem analysis: Identify and analyses well-defined engineering problems using codified standard methods.

PO3. Design/ development of solutions: Design solutions for well-defined technical problems and assist with the design of systems components or processes to meet specified needs.

PO4. Engineering Tools, Experimentation and Testing: Apply modern engineering tools and appropriate technique to conduct standard tests and measurements.

PO5. Engineering practices for society, sustainability and environment: Apply appropriate technology in context of society, sustainability, environment and ethical practices.

PO6. Project Management: Use engineering management principles individually, as a team member or a leader to manage projects and effectively communicate about well-defined engineering activities.

PO7. Life-long learning: Ability to analyse individual needs and engage in updating in the context of technological changes.

COURSE OUTCOMES (CO)

By the end of the course the students are expected to learn:

CO-1: Prepare plans using Plane Table Surveys

CO-2: Prepare plans using Theodolite Survey

CO-3: Find distances and elevations using Tacheometer

CO-4: Prepare plans using Total Station instrument

CO-5: Locatecoordinates ofstations using GPS

Mapping of Course Outcomes with Programme Outcomes to be done according to the matrix given below:

Course Outcomes	EXPECTED MAPPING WITH PROGRAMME OUTCOMES (1- Weak Correlation; 2- Medium correlation; 3- Strong Correlation)						
	PO-1	PO-2	PO-3	PO-4	PO-5	PO-6	PO-7
CO-1	3	2	2	3	1	2	3
CO-2	3	3	3	3	1	2	3
CO-3	3	3	3	3	1	2	3
CO-4	3	2	2	3	1	2	3
CO-5	3	2	2	3	1	2	3

GUIDELINES FOR TEACHERS

To implement Outcome Based Education (OBE) knowledge level and skill set of the students should be enhanced. Teachers should take a major responsibility for the proper implementation of OBE. Some of the responsibilities (not limited to) for the teachers in OBE system may be as follows:

- Within reasonable constraint, they should manoeuvre time to the best advantage of all students.
- They should assess the students only upon certain defined criterion without considering any other potential ineligibility to discriminate them.
- They should try to grow the learning abilities of the students to a certain level before they leave the institute.
- They should try to ensure that all the students are equipped with the quality knowledge as well as competence after they finish their education.
- They should always encourage the students to develop their ultimate performance capabilities.
- They should facilitate and encourage group work and team work to consolidate newer approach.
- They should follow Blooms taxonomy in every part of the assessment.

Bloom's Taxonomy

Level	Teacher should Check	Student should be able to	Possible Mode of Assessment
Creating	Students ability to create	Design or Create	Mini project
Evaluating	Students ability to Justify	Argue or Defend	Assignment
Analysing	Students ability to distinguish	Differentiate or Distinguish	Project/Lab Methodology
Applying	Students ability to use information	Operate or Demonstrate	Technical Presentation/ Demonstration
Understanding	Students ability to explain the ideas	Explain or Classify	Presentation/Seminar
Remembering	Students ability to recall (or remember)	Define or Recall	Quiz

GUIDELINES FOR STUDENTS

Students should take equal responsibility for implementing the OBE. Some of the responsibilities (not limited to) for the students in OBE system are as follows:

- Students should be well aware of each UO before the start of a unit in each and every course.
- Students should be well aware of each CO before the start of the course.
- Students should be well aware of each PO before the start of the programme.
- Students should think critically and reasonably with proper reflection and action.
- Learning of the students should be connected and integrated with practical and real life consequences.
- Students should be well aware of their competency at every level of OBE.

ABBREVIATIONS AND SYMBOLS

List of Abbreviations

General Terms			
Abbreviations	**Full form**	**Abbreviations**	**Full form**
N-S	North – South	W.C.B	Whole Circle Bearing
LOS	Line of Sight	R.B.	Reduced Bearing
LOC	Line of Collimation	P.C.	Point of Curve
L	Latitude	P.T.	Point of Tangency
D	Departure	P.I.	Point of Intersection
s	Staff Intercept	EDM	Electromagnetic Distance Meter/ Measurement
f	Focal Length of the objective		
GIS	Geographical Information System	CAD	Computer Added Design
		RS	Remote Sensing
GPS	Global Positioning system	TS	Total Station
LED	Liquid Crystal Diode	EM	Electromagnetic
LCD	Light Emitting Diode		

List of Units and Symbols

Symbols	Description	Symbols	Description
α (R)	Right Deflection angle	Λ	Weblength
β (L)	Left Deflection angle	Φ	Phase Angle of web
Δ	Deflection Angle	N	Number of Sides
k	Multiplying Constant in Tacheometer	E	Closing Error in Traverse
Da, Dc	Degree of Curve	I	Stadia Hair Interval
f	Focal Length	C	Additive Constant in Tacheometer

LIST OF FIGURES

CONTENTS

1 Plane Table Surveying

UNIT SPECIFICS

Through this unit we have discussed the following aspects:

- Principle of plane table surveying
- Instruments used in plane table surveying
- Working operations of plane table surveying
- Methods of plane table surveying
- Merits and demerits of plane table surveying

RATIONALE

This unit describes purpose, principle, and methodologies of plane table surveying. The instruments used in the plane table survey and their uses are stated with neat diagrams. This unit discusses working operations of plane table survey including fixing of table, levelling, centring and orientation of table and sighting of objects.

The survey by the plane table is done for locating various ground feature and plane table stations and plotting them on the sheet. The field observations and plotting on sheet are done simultaneously in this survey. The methods of plane table survey include radiation, intersection, traversing and resection. Nowadays the utility of plane table survey is limited with invention of electronic distance measurement methods and plotting/drawing software.

PRE-REQUISITE

Basic Surveying (Third Semester Diploma)

UNIT OUTCOMES

List of outcomes of this unit is as follows:

U1-O1: State the principle of plane table survey
UI-O2: Describe the instruments used in plane table survey
U1-O3: Explain the operations of plane table survey
U1-O4: Apply methods of plane table survey
U1-O5: List the merits and demerits of plane table survey

Unit-1 Outcomes	EXPECTED MAPPING WITH COURSE OUTCOMES (1- Weak Correlation; 2- Medium correlation; 3- Strong Correlation)				
	CO-1	CO-2	CO-3	CO-4	CO-5
U1-O1	2	1	1	1	1
U1-O2	3	1	1	1	-
U1-O3	3	1	1	1	-
U1-O4	3	2	1	1	-
U1-O5	3	-	-	-	-

1.1 INTRODUCTION

The plane table surveying is a graphical method of surveying in which ground field work and plotting on sheet are done simultaneously. The beauty of this survey is that it omits recording of data on the field and using it for plotting on sheet. This survey is suitable for small to medium size of field. It is the most useful in magnetic areas where compass survey does not work.

1.2 PRINCIPLE OF PLANE TABLE SURVEYING

The principle of plane table survey is based on parallelism. It states

(i) A line joining the points on the plane table are made to lie parallel to its corresponding line joining the ground stations during working at each station.

(ii) The plane table at each station must be placed identical, i.e. at each ground station the table must be oriented in some fixed direction.

1.3 INSTRUMENTS USED IN THE PLANE TABLE SURVEYING

The instruments used are as follows:

1.3.1 The plane table

It is a wooden table mounted on tripod. It may be rotated about its vertical axis and may be fixed at any position. The sizes of plan table available are 600 mm x 500 mm, 750 mm x 600 mm or 100 mm x750 mm. The height of tripod used is usually 1200 mm. (Fig. 1.1)

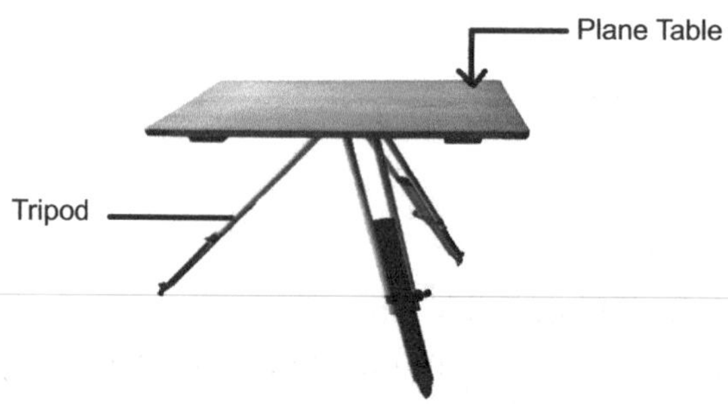

Fig. 1.1: Plane Table

1.3.2 Alidade

The Alidade may be two types.

(i) Plain Alidade

(ii) Telescopic Alidade

The plain Alidade consists of a wooden or metal rule has two edges. One straight and other one is bevelled. The straight edge is made of gunmetal or brass and used as a ruler. The bevelled edge is known the fiducial edge. It has two vanes at the both ends. The vanes are hinged at the ends and may be folded when the alidade is not in use. One of the vanes is the eye vane or the sight vane provided with narrow silt. The other vane is called as object vane. It is open and carries a hair or thin wire. The surveyor looks through the narrow silt towards the object or station coinciding the hair of object vane and establishes a horizontal line of sight parallel to the ruler. The problem with a simple alidade is that it can be used only to take horizontal sight to the objects at alidade level. (Fig. 1.2)

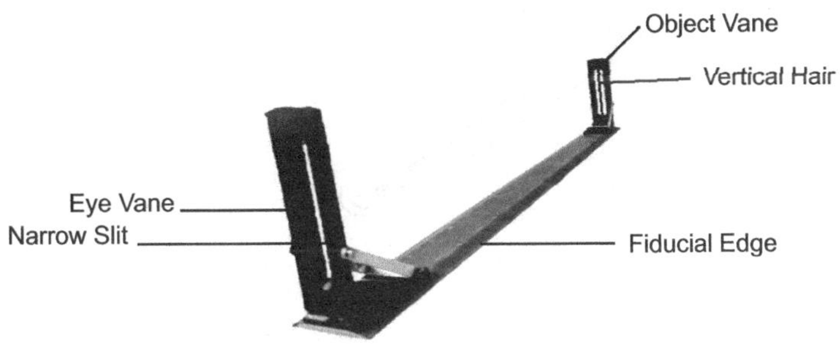

Fig. 1.2: Plain Alidade

Telescopic alidade is used to take inclined sights to the objects at above or below plane table level. It is used to enhance the accuracy of the sight taken. It consists of a wooden or metal rule provided with a spirit level tube and a telescope so that the alidade can be levelled with the working station and telescope provides inclined line of sight. A scale is marked on the horizontal axis and lines are drawn along the straight ruler. It is also mounted with a vertical circle that measures the angle of the object with the horizontal axis. (Fig. 1.3)

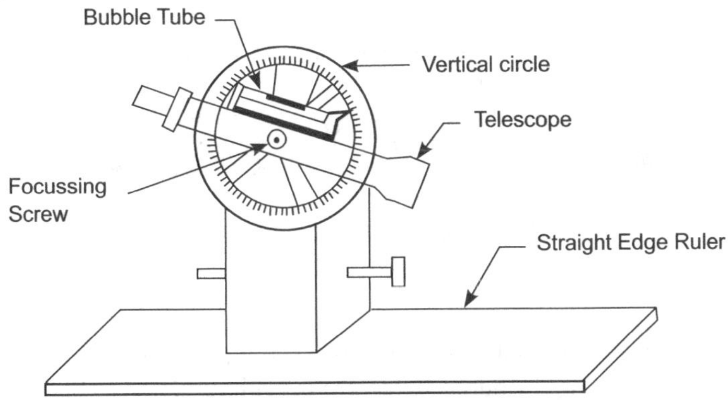

Fig. 1.3: Telescopic Alidade

1.3.3 Plumbing Fork and Plumb Bob

Plumbing fork is used for centring the table over the station occupied by the plane table when the position of that station is already plotted on the sheet. In other words, it is used for transferring the ground point on the sheet so that the plotted point and ground station are in same vertical line. It is a U-shaped metal frame that has two different arms. Upper arm is horizontal and the lower is inclined at a certain angle. The upper arm is provided with a pointer at the end while the lower arm has a hook at the end, from which plumb bob is suspended. The upper arm is kept on the sheet. The table is so placed that plumb bob touches the ground station, then pointer of upper arm defines plotted position of the ground station. (Fig. 1.4).

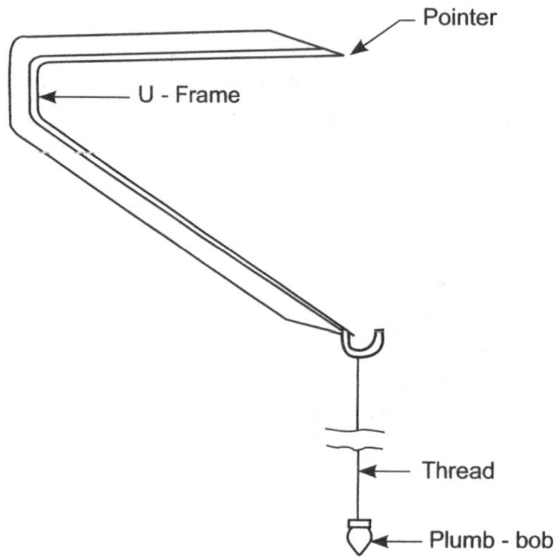

Fig. 1.4: Plumbing fork with U- frame

1.3.4 Trough Compass

A trough compass is used to orient the table to magnetic north. A trough compass consists of a long, narrow rectangular box with a long narrow magnetic needle mounted on a pivot. The long sides are used as ruler to draw a arrow along magnetic north. (Fig. 1.5)

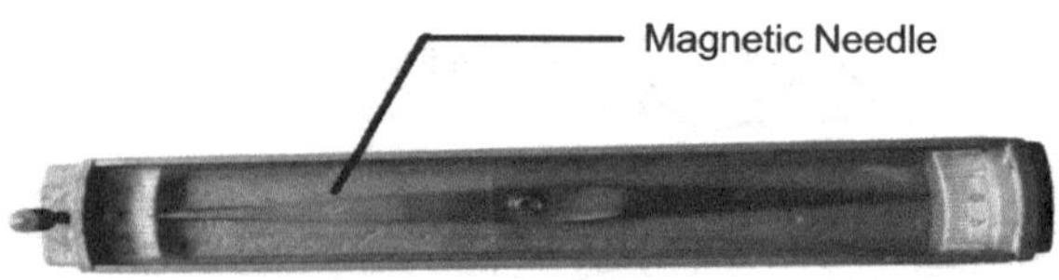

Fig. 1.5: Trough Compass

1.3.5 Spirit Level

Spirit level is used to level the table. The table is levelled by keeping the level on the board in two mutually perpendicular directions and getting the bubble in centre in both directions. (Fig. 1.6)

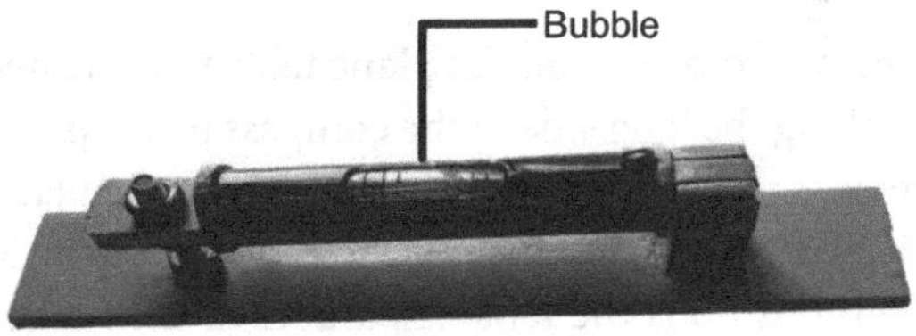

Fig. 1.6: Spirit Level

1.4 WORKING OPERATIONS

Following operations are carried out during plane table survey

1. Fixing the table on the tripod
2. Setting out the table

   It consists of three activities

   (a) Levelling the table
   (b) Centring
   (c) Orientation
3. Sighting the stations/objects

1.4.1 Fixing the table on the tripod

The tripod is unfolded, and legs are spread so that its height is approximately 1.2 metre above the ground level. The plane table is placed on tripod top and clamp screw is tightened into brass annular ring of the plane table.

1.4.2 Setting out of the table

It consists of three activities

(a) Levelling: The table is levelled by putting the level on the table in two mutual perpendicular direction and getting the bubble in centre in both directions by tilting of board and adjusting the legs of the tripod.

(b) Centring: The table is so placed over the station on the ground that the point plotted on the sheet corresponding to the station occupied must be exactly over the station on the ground. This is called centring of the plane table. The centring is done by plumbing fork, the procedure is described earlier in instruments section.

(c) Orientation: Orientation is a process of putting the plane table into the fixed direction so that a line on the plane table and corresponding line on ground show the same direction. This is a necessary when more than one instrument station are used. If the orientation is not done, the table will not be parallel to itself at different positions, that will result in a complete distortion of the map.

Main methods of orientation

(i) Orientation by trough compass
(ii) Orientation by back sighting.

(i) Orientation by trough compass

In this method, the trough compass is so placed on the plane table that the needle freely floats and rests in N-S direction. A line is drawn along the long side of the compass box. This line shows magnetic north. When the table is to be oriented on another station, the compass is to be placed along this line showing magnetic north and table is oriented by rotating it until the needle rests in N-S direction. The table is then fixed in the position. In both positions the table is parallel to each other. This method is fast but less accurate.

(ii) Orientation by back sighting

Let the table is placed at any fixed direction at station A, represented by a point 'a' on sheet. Now the table is to be oriented in same direction at station B. Put the table on station A and draw a line ab towards station B with the help of alidade, point 'b' shows station B. Place the table on station B and keep the alidade along line ba and rotate the table in such a way that the line of sight passes through the station A. In this position ground line AB will coincide with plotted line ab and table thus oriented and is clamped in this position.

1.4.3 Sighting the stations/objects

After setting of the table i.e. completing of levelling, centring and orientation , the objects or points to be located are sighted by the alidade. Keeping alidade pivoted about the plotted location of the plane table station and turning the alidade in such a way that the line of sight bisects the object or signal at the point to be plotted. A line is drawn along the ruler of the alidade from the plotted position of the station.

1.5 METHODS OF PLANE TABLE SURVEYING

There are four methods of plane table surveying

1. Radiation Method
2. Intersection Method
3. Traversing Method
4. Resection Method

First two methods used for locating the details while last two methods are more accurate, used for locating plane table stations. Here first three methods are described.

1.5.1 Radiation Method

This method of plane table surveying is suitable for locating the details from single station and when distances of objects are small. The direction of the objects or the points to be located are indicated by drawing radial lines by sighting the objects or the points with the help of alidade. The horizontal distance between instrument and stations are then measured and stations are located on the sheet to some scale.

Procedure:

1. Set up the plane table at a station X, after levelling, transfer the ground point X on the sheet by plumbing fork as 'x' .
2. Keeping alidade pivoted at x, sight object A. Draw the ray (xa) along fiducial edge of the alidade. Similarly draw the rays towards stations B, C, D, E etc. by sighting them.

3. Measure ground distances XA, XB, XC, XD etc. and plot them to some scale along their corresponding rays (xa, xb, xc, xd, etc.) and mark points a, b, c, d, etc.

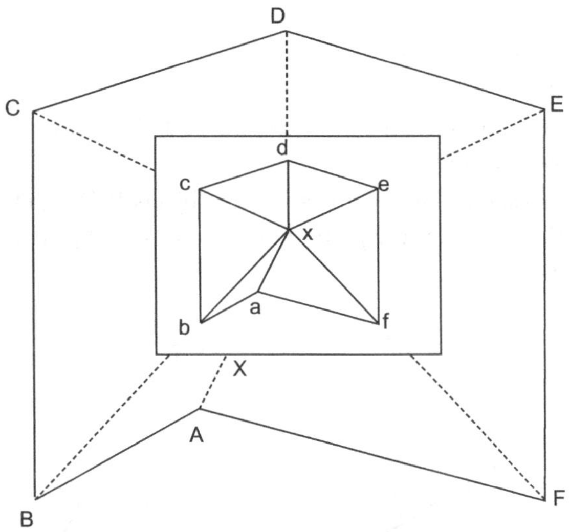

Fig. 1.7: Radiation Method

Table 1.1: Radiation Method

Instrument at	Alidade pivoted about	Sight to	Draw the resector	Remarks
X	x	A	xa	Mark a, b, c, d, e to the scale as per ground distances and join a, b, c, d, e, f
X	x	B	xb	
X	x	C	xc	
X	x	D	xd	
X	x	E	xe	
X	x	F	xf	

1.5.2 Intersection method

The intersection method is suitable when distances of objects to be located, are large or cannot be measured properly due to ground conditions. This method is preferred in small survey work and for mountainous regions. The objects are located by sighting them from two plane table stations whose positions are already plotted and rays are drawn towards the objects. The intersection of rays towards the objects from two different plane table stations determines the plotted position of the objects.

Procedure:

1. Set the table at station A, after levelling, transfer the ground point A on the sheet as 'a' with the help of plumbing fork then clamp the table.
2. Mark the north direction on the paper using trough compass.

3. Keeping alidade pivoted at a, sight it towards station B and a ray is drawn. By measuring ground distance AB, plot along the ray to some scale and mark point b corresponding station B thus line ab drawn.
4. Keeping alidade pivoted at a, sight objects P, Q, R, etc. and corresponding rays are drawn.
5. Shift the table at station B .
6. After levelling, orient it by back sighting A. Keeping alidade pivoted at b, sight object P, Q, R, etc. and corresponding rays are drawn.
7. The intersection of corresponding rays towards the objects P, Q, R, etc. from two different plane table stations A and B determines the plotted position of the objects P, Q, R, etc., thus p, q, r, etc. are drawn on the paper.

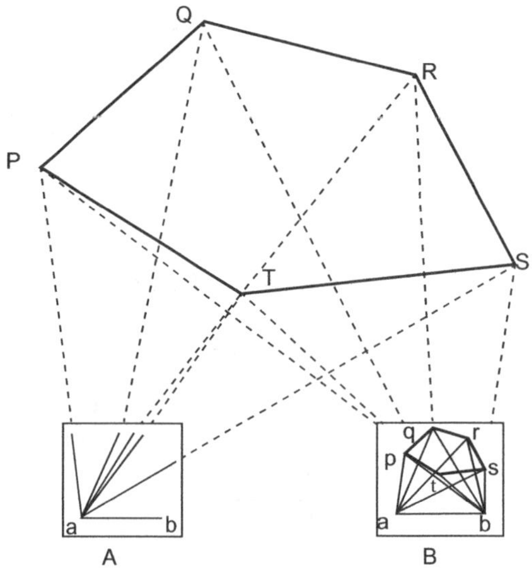

Fig. 1.8: Intersection Method

Table 1.2: Intersection Method

Instrument at	Alidade pivoted about	Sight to	Draw the resector	Remarks
A	A	P	ap	
A	A	Q	aq	
A	A	R	ar	
A	A	S	as	
A	A	T	at	
B	B	A	-	Orientation
B	B	P	bp	p, q, r, s, t are marked by intersection of resectors from stations A and B.
B	B	Q	Bq	
B	B	R	Br	
B	B	S	Bs	
B	B	T	Bt	

1.5.3 Traversing Method

The method of traversing is used when the stations have not been plotted previously. In this method, traverse stations are first selected then the stations are plotted by method of radiation by taking back sight on the previous station and a fore sight to the next station. The location of stations are plotted by measuring the distance between two stations as done in radiation method.

Procedure:

1. Set the table at station A, after levelling ,transfer the ground point A on the paper as 'a' with the help of plumbing fork .
2. Mark the north direction on the paper using trough compass.
3. Keeping alidade pivoted at a, sight it towards station B and a ray is drawn. By measuring ground distance AB, plot along the ray to some scale and mark point b corresponding station B thus line ab is drawn. Similarly draw the ray towards E, measure AE and thus line ae is drawn
4. Shift the table at station B. After levelling, orient it by back sighting A.
5. Keeping alidade pivoted at b, sight it towards station C and a ray is drawn. By measuring ground distance BC, plot along the ray to some scale and mark point c corresponding station C thus line bc is drawn. Similarly, the table is shifted at other stations and the traverse is completed.

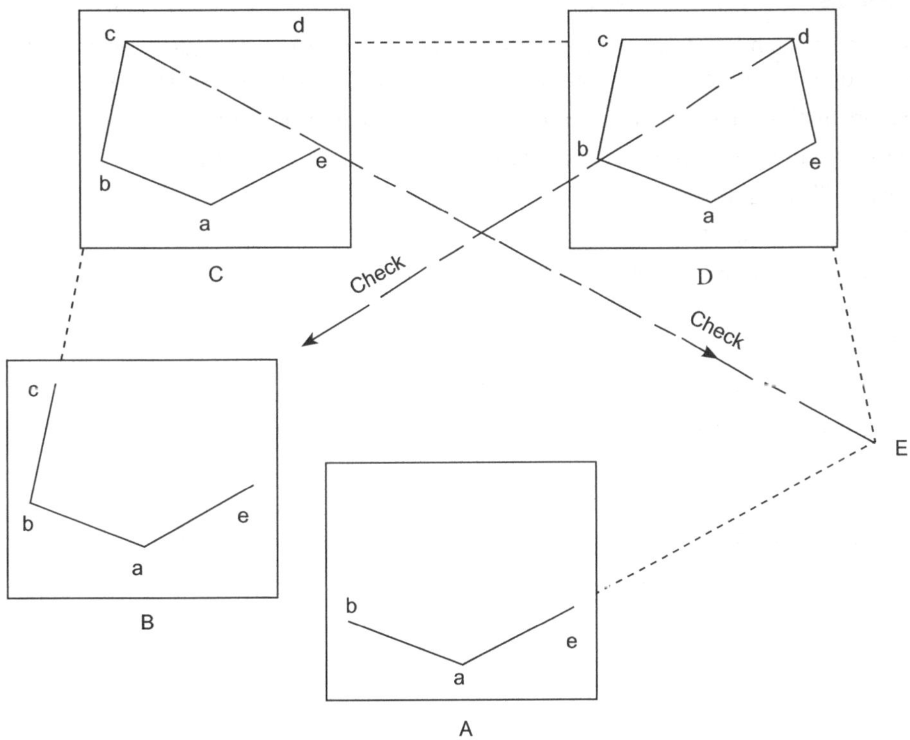

Fig. 1.9: Traversing Method

Table 1.3: Traversing Method

Instrument at	Alidade pivoted about	Sight to	Draw the resector	Remarks
A	a	B	ab	
A	a	E	ae	
B	ba	A	-	Orientation
B	b	C	bc	
C	cb	B	-	Orientation
C	c	D	cd	
C	c, e	E	-	Check
D	dc	C	-	Orientation
D	b	B	-	Check
D	E	E	de	

1.6 MERITS AND DEMERITS OF PLANE TABLE SURVEYING

1.6.1 Merits

1. The observation and plotting are done simultaneously. All ground features remain before the eyes of surveyor, hence there is no possibilities of omitting necessary measurements.
2. The errors in plotting the details can be checked by drawing checklines
3. As the area is in view, irregular objects can be plotted with great accuracy
4. No great skill is required
5. It is cheaper than theodolite survey
6. It is the most useful in magnetic areas where compass survey does not work.

1.6.2 Demerits

1. It is not suitable in rainy season and densely wooded areas
2. As field measurements are not recorded, it is highly inconvenient if the survey is to be plotted to some different scale.
3. It is not suitable for accurate work.
4. The apparatus with accessories is inconvenient to carry due to heaviness
5. There are many small accessories, hence likely to be lost.

UNIT SUMMARY

- Principle of plan table survey is based on parallelism.
- Plan table survey is graphical method in which plotting, and field work are done simultaneously.
- Sizes of plane table usually available are 600 mm × 500 mm, 750 mm × 600 mm and 1000 mm × 750 mm.

- Purpose of the plane table accessories
  - (i) Plane table – Used for fixing the drawing sheet.
  - (ii) Tripod – Used to support the plane table at required height
  - (iii) Trough Compass – To mark the direction on drawing sheet
  - (iv) Alidade – Used to draw the line of sight
  - (v) Spirit Level – used to check the levelling of plane table
  - (vi) Plumb bob- used to mark the centre point on the ground.
  - (vii) Plumbing-fork – Used to transfer the ground point on drawing sheet.
- Plane table survey is most suitable for small scale map.
- It is recommended in magnetic areas because there is no effect of magnetic field.
- Limitations: It is not suitable in rainy season.
- Methods of plan table
  - (i) Radiation
  - (ii) Intersection
  - (iii) Traversing
  - (iv) Resection

EXERCISE

Multiple Choice Questions

1. Plan table survey is based on the method of ________.

   (a) Ranging (b) Contouring
   (c) Traversing (d) Triangulation

2. U- Frame is used for

   (a) For Focusing (b) For Levelling
   (c) For Centring (d) For Orientation

3. What is the purpose of the alidade in plane table survey?

   (a) For Sighting (b) For Levelling
   (c) For transferring the point to ground (d) All the above

4. What is the advantage of plane table survey?

   (a) Accurate output (b) Suitable for wet climate
   (c) Used in magnetic areas (d) Suitable

5. Following method is not the example of plane table survey.

   (a) Radiation (b) Intersection
   (c) Trisection (d) Traversing

6. Following method of plane table survey requires two instruments station.
   (a) Traversing (b) Radiation
   (c) Intersection (d) Resection
7. Following methods of plane table are used to locating the details of the survey
   (a) Resection, Intersection (b) Radiation, Intersection
   (c) Radiation, Resection (d) Traversing, Resection
8. In the magnetic area which type of survey can be preferred
   (a) Compass Survey (b) Theodolite Survey
   (c) Plan table survey (d) All the above
9. Plane table survey is more suitable for wet climate
   (a) True (b) False
10. Error occurred due to the orientation in plane table survey can be checked by
    (a) Calculating area (b) Measuring Bearings
    (c) Calculating Volume (d) Measuring Angles
11. How many methods of plane table survey are there?
    (a) 3 (b) 1
    (c) 4 (d) 2
12. Which instrument is used for levelling in a plane table survey?
    (a) U- Frame (b) Compass
    (c) Plumb bob (d) Spirit level
13. In plane table surveying, plotting and observations are done simultaneously.
    (a) True (b) False
14. The plane table survey is the ________ method.
    (a) mathematical (b) graphical
    (c) analytical (d) none of above
15. Trough compass is used for
    (a) Centring (b) Orientation
    (c) Levelling (d) Sighting

ANSWERS															
1.	(c)	2.	(c)	3.	(a)	4.	(c)	5.	(c)	6.	(c)	7.	(b)	8.	(c)
9.	(b)	10.	(d)	11.	(c)	12.	(d)	13.	(a)	14.	(b)	15.	(b)		

Short and Long Answer Type Questions

Short Questions:

1. Explain briefly, plane table survey.
2. Explain the principle of plane table survey.

3. What is the difference between simple alidade and telescopic alidade?
4. Write the advantages and disadvantages of the plane table survey.
5. Explain the orientation of the plane table survey.

Long Questions:

1. Describe the following methods of plane table surveying:
   (i) Radiation
   (ii) Intersection
   (iii) Traversing
2. Explain each accessory used in plane table survey.
3. Discuss the working operations of plane table survey.
4. Discuss the differences between radiation and Intersection methods.

PRACTICALS

Experiment No. - 1

Objective: Use plane table survey to prepare plans of a plot of seven-sided closed traverse by Radiation Method.

Required Accessories:

1. Drawing Board
2. Tripod stand
3. Alidade
4. Spirit Level
5. Trough Compass
6. Plumbing fork with plumb- bob
7. Drawing Sheet
8. Tape for distance measurement
9. Pencil, Eraser, Clamp or Scotch tape etc.

Procedure:

- Fix the plane table on tripod at station 'O' and paste the drawing sheet on plane table.
- Level the plane table using the spirit level by tilting the board and adjusting the legs of tripod.
- Transfer the ground station 'O' on the drawing sheet as 'o' with the help of plumbing fork using plumb bob.
- Mark the north direction on the drawing sheet with the help of trough compass.
- Pivot the alidade on 'o' and draw the line of sight towards objects A, B, C, D, E, F and G by sighting with the help of alidade.
- Measure the distances OA, OB, OC, OD, OE, OF and OG and mark the points on drawing sheet to some scale as a, b, c, d, e, f and g respectively.
- Join the points to complete traverse abcdefga.

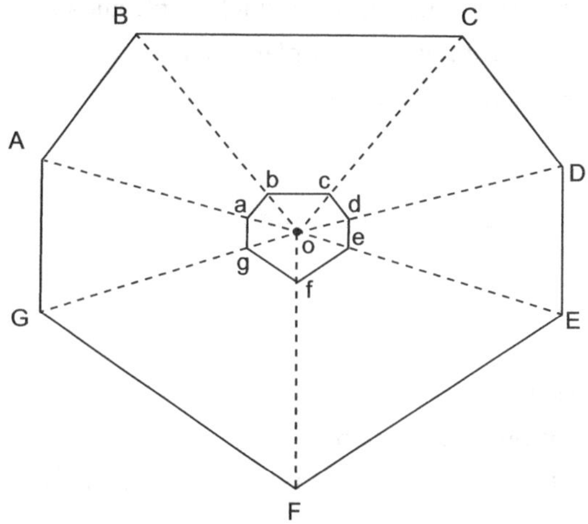

Fig. 1.10: Radiation Method – Seven-sided closed traverse

Experiment No. – 2

Objective: Use plane table survey to prepare plans, locate details by Intersection Method.

Required Accessories:

1. Drawing Board
2. Tripod stand
3. Alidade
4. Spirit Level
5. Trough Compass
6. Plumbing fork with plumb- bob
7. Drawing Sheet
8. Tape for distance measurement
9. Pencil, Eraser, Clamp or Scotch tape, etc.

Procedure:

- Fix the plane table on tripod at station 'P' and paste the drawing sheet on plane table.
- Level the plane table using the spirit level by tilting the board and adjusting the legs of tripod.
- Transfer the ground station 'P' on the drawing sheet as 'p' with the help of plumbing fork.
- Select another station Q and draw a ray from P towards Q by sighting with the help of alidade. Measure the distance PQ on ground and to some scale mark the point q.
- Draw the rays from A towards objects X_1, X_2, X_3, X_4 and X_5 by sighting with the help of alidade.
- Set the plane table on station Q such that station Q is exactly below the sheet point q using plumbing fork. Level the table and orient it by back sighting P.
- Draw the rays from Q towards object X_1, X_2, X_3, X_4 and X_5 by sighting with the help of alidade.
- The intersection of corresponding rays towards the object X_1, X_2, X_3, X_4 and X_5 from two different plane table stations P and Q determines the position of corners of plan x_1, x_2, x_3, x_4 and x_5.
- Join x_1, x_2, x_3, x_4 and x_5. to complete the plan x_1, x_2, x_3, x_4, x_5.

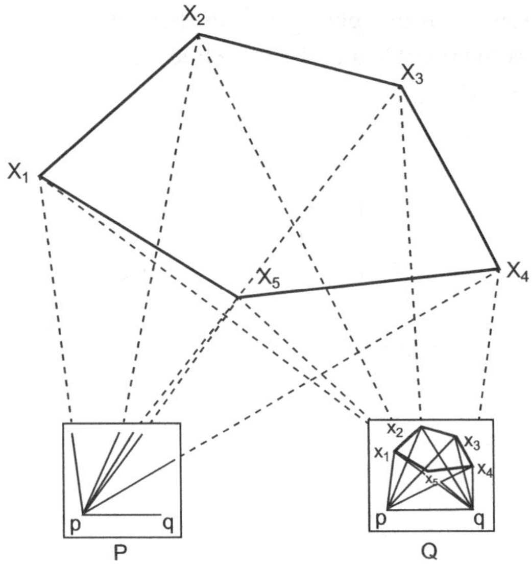

Fig. 1.11: Locating Details by Intersection Method

Experiment No. – 3

Objective: Use plane table survey to prepare plans, locate details by Traversing Method.

Required Accessories:

1. Drawing Board
2. Tripod stand
3. Alidade
4. Spirit Level
5. Trough Compass
6. Plumbing fork with plumb- bob
7. Drawing Sheet
8. Tape for distance measurement
9. Pencil, Eraser, Clamp or Scotch tape etc.

Procedure:

- Fix the plane table on tripod at station 'P' and paste the drawing sheet on plane table.
- Level the plane table using the spirit level by tilting the board and adjusting the legs of tripod.
- Transfer the ground point P on the sheet as 'p' with the help of plumbing fork.
- Mark the north direction on the sheet using trough compass.
- Draw the ray from P towards Q by sighting with the help of alidade. Measure ground distance PQ and to some scale mark the point 'q' and draw the line pq. Similarly draw the ray towards T, measure PT and to some scale mark the point 't' and draw the line pt.
- Shift the table at station Q. After levelling, orient it by back sighting P.

- Draw the ray from Q towards R by sighting with the help of alidade. Measure ground distance QR and to some scale mark the point 'r' and draw the line qr.
- Similarly, the table is shifted at other stations and the traverse is completed.

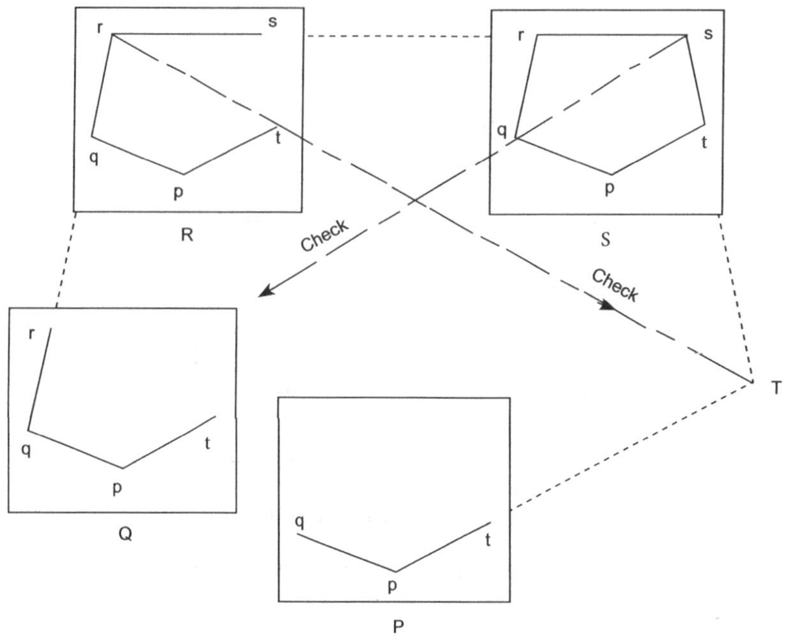

Fig. 1.12: Locating Details by Traversing Method

KNOW MORE

Resection Method of Plane Table survey

The fourth method of plane table survey is resection method, which is more accurate method than other methods of plane table surveying. It is used for locating the plane stations.

Resection is the method of determining the plotted position of the station which is occupied by the plane table by sighting known stations whom locations have already been plotted.

The resection problem can be solved by following methods.

1. Resection after orientation by compass
2. Resection after orientation by back sighting
3. Resection by three-point problem
4. Resection by two-point problem

Errors in Plane Table Survey

(i) Instrumental Error

  1. Error due to inclined plane table
  2. Error due to fiducial edge might not be straight

3. Plane table is not properly tight with tripod.
4. Needle of trough compass is not properly balanced.
5. Error due to imperfection of Spirit level
6. Object vane and eye vane may not be vertical

(ii) Errors in Operations

1. Plane table may not be perfectly levelled.
2. Inaccurate centring of plane table
3. Inaccurate orientation of plane table
4. Imperfect bisection of objects

(iii) Plotting errors

1. Taking wrong scale by mistake
2. Incorrect linear measurement

Types of Plane Table

Following three types of plane table are used.

1. **Traverse Table**: It is the ordinary plane table, which can be mounted on tripod stand by tightening clamp screw of tripod into brass annular ring of the plane table. It is levelled by adjusting the legs of the tripod.
2. **Johnsons Table:** It is provided with additional ball and Socket arrangement for levelling purpose.
3. **Coast Survey table:** It is high quality table provided with three levelling screw arrangement similar to the levelling instruments.

Dynamic QR code for Resection method

Resection method of plane table surveying

REFERENCES & SUGGESTED READINGS

1. B.C Punmia, Surveying Vol –II, Laxmi Publication (P) Ltd, 2023.
2. N.N. Basak, surveying & Levelling,McGraw hill Iindia, Private Limited, Noida, 2017.
3. Bhavikatti SS, Surveying and Levelling Vol. II, I.K International Publishing House Pvt. Ltd, 2016.
4. R. Agor, A Text Book of Surveying & Levelling, Khanna Publication, 2015.
5. S.K. Duggal, Surveying Vol. I, McGraw Hill Publishing Company Ltd., 2013.
6. C. Venkatramaiah, Textbook of Surveying, University Press (India) Limited, 2011.
7. Saikia MD, Das BM, Das MM, Surveying, Prentice Hall India Learning Limited, 2010.
8. T.P. Kanetkar and S.V. Kulkarini, Surveying and Levelling, Vidyarthi Griha Prakashan, 1967.
9. NPTEL video link - https://youtu.be/vT_7OmzFiSE
10. Notes Link- https://www.madeeasy.in/Uploads/examsolution/Surveying.pdf

Dynamic QR Code for Further Reading

Following QR codes are given for further study of plan table surveying.

1. Dynamic QR code for plan table survey (Theory)

Introduction to Plane table survey

2. Dynamic QR code for Accessories used in plane table (Video)

Accessories used in plane table surveying

3. Dynamic QR Code for Radiation method (Video)

Rediation method of plane table surveying

4. Dynamic QR Code for Intersection method (Video)

Intersection method of plane table surveying

5. Dynamic QR Code for Traversing method (Video)

Traversing method of plane table surveying

2 Theodolite Surveying

UNIT SPECIFICS

Through this unit we have discussed the following aspects:

- Purpose and types of Theodolites
- Different parts of transit Theodolite and their function
- Working of transit Theodolite
- Temporary adjustment of Theodolite
- Measurement of horizontal and vertical angle by transit Theodolite
- Some field applications of Theodolite
- Traversing by Theodolite
- Traverse Computations

RATIONALE

This unit describes the purpose and types of Theodolites. The Theodolite is most precise instrument used to determine horizontal and vertical angle. This chapter presents various parts of transit theodolite with their functions. The terminology used in Theodolite including transiting, face left, face right and fundamental axes of transit theodolite are defined clearly . The temporary adjustment of Theodolite is necessary before taking observations. The direct and repetition methods of measurement of horizontal angle and measurement of vertical angle by Theodolite are discussed in this chapter.

The various uses of theodolite including measurement of magnetic bearing of line, prolonging, and ranging a line and measurement of deflection angle are dealt in this chapter. This chapter also discusses Theodolite traversing which includes checks for open and closed traverse, calculation of bearing from angles. Traverse computation including determination of latitude, departure, consecutive & independent coordinates and balancing the traverse by Bowditch's rule and transit rule, is also discussed. The Gale's traverse table is also presented.

PRE-REQUISITES

Mathematics: Basic knowledge of Trigonometry

Basic Surveying (Third Semester Diploma)

UNIT OUTCOMES

List of outcomes of this unit is as follows:

U2-O1: State the uses and types of Theodolites
U2-O2: Explain the working of Theodolite
U2-O3: Determine horizontal and vertical angle by using Theodolite
U2-O4: Perform field work using Theodolite
U2-O5: Apply methods of traversing using Theodolite

Unit-2 Outcomes	EXPECTED MAPPING WITH COURSE OUTCOMES (1- Weak Correlation; 2- Medium correlation; 3- Strong Correlation)				
	CO-1	CO-2	CO-3	CO-4	CO-5
U2-O1	1	3	2	2	1
U2-O2	1	3	2	2	1
U2-O3	1	3	2	2	1
U2-O4	1	3	2	2	1
U2-O5	1	3	2	2	1

2.1 INTRODUCTION

The Theodolite is the most versatile and precise instrument widely used for the measurement of horizontal and vertical angle. It has wide application in the field such as locating points on a line, laying grades, finding difference in elevation, setting out curves, prolonging survey lines etc.

In this chapter vernier theodolite is discussed as the basic features of a Theodolite may be illustrated conveniently by simplified diagram of a vernier Theodolite although Digital Theodolites are more accurate, compact and convenient and used nowadays.

2.2 CLASSIFICATION OF THEODOLITE

Theodolites may be classified two ways

2.2.1 Classification based on Movement of Telescope

(i) Transit Theodolite
(ii) Non-transit Theodolite

Transit Theodolite

In transit Theodolite, the telescope can be rotated through 180 0 in the vertical plane about its horizontal axis i.e. line of sight can be reversed. A transit theodolite is simply called 'transit'. It is widely used.

Non-transit Theodolite

In Non-transit Theodolite, the telescope cannot be transited, i.e., cannot be resolved through 180 0. The non-transit theodolites have now become obsolete.

2.2.2 Classification based on Arrangement of Reading Observations

(i) Vernier Theodolite
(ii) Micrometer Theodolite
(iii) Optical Theodolite
(iv) Electronic or Digital Theodolite

Vernier Theodolite

In this type of Theodolite, vernier is used for reading horizontal and vertical graduated circles. The least count of vernier Theodolite is 20".

Micrometer Theodolite

In this type of Theodolite, micrometre is used for reading horizontal and vertical graduated circles. The least count of micrometer Theodolite is 1".

Optical Theodolite

In optical theodolite, graduated glass circle is used, and the reading is reflected to observer eye by means of a system of prisms and lenses.

Electronic or Digital Theodolite

In Electronic or Digital Theodolite, the reading of angle is found in digital form. These types of theodolites are accurate and more convenient, hence other types of Theodolites have become obsolete.

2.3 COMPONENTS OF TRANSIT THEODOLITE

The transit theodolite consists of following components (see Fig. 2.1)

1. Telescope
2. Vertical clamp
3. Arm of the vertical circle clamp
4. Standard
5. Line of sight
6. Upper clamp
7. Axis of plate bubble
8. Upper plate
9. Lower plate
10. Lower clamp
11. Upper Tribrach plate
12. Foot screw
13. Lower Tribrach plate
14. Tripod top
15. Plumb bob
16. Vertical axis
17. Tripod clamping screw
18. Levelling head
19. Horizontal graduated arc
20. Plate level
21. Vernier frame
22. Horizontal or Trunnion axis
23. Altitude bubble
24. Vertical circle

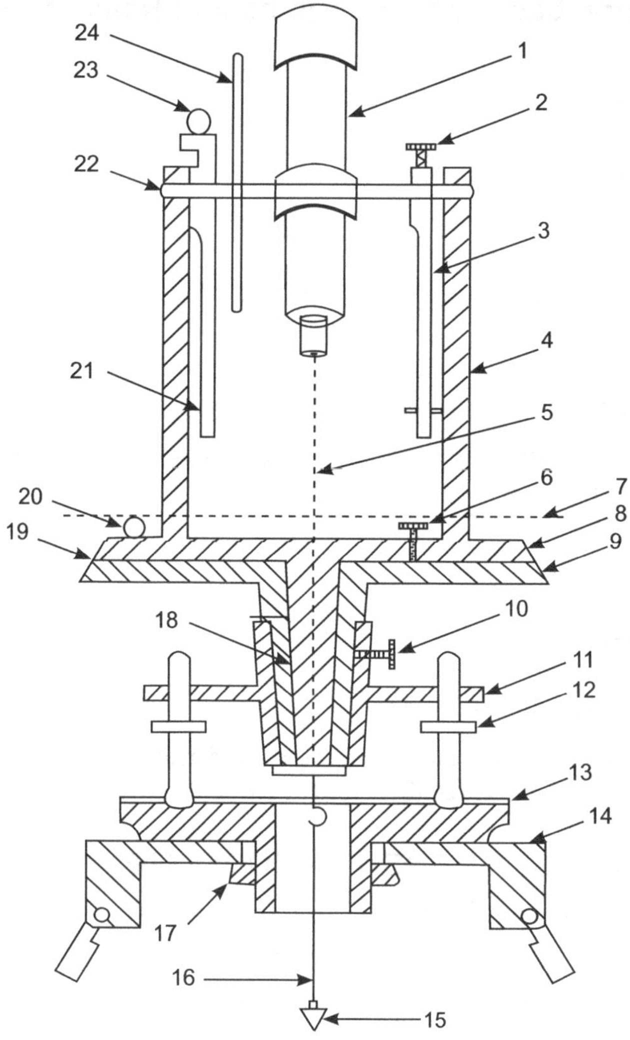

Fig. 2.1: Components of transit theodolite

2.3.1 The Telescope

The telescope (1) is mounted on a spindle. The spindle coincides with horizontal axis, it is also called trunnion axis (22). The telescope provides horizontal as well as inclined line of sight (5). Generally internal focusing telescope is used.

2.3.2 The Vertical Circle

The vertical circle (24) is a graduated circular arc fixed with telescope. The centre of circular arc coincides with trunnion axis, the vertical circle rotates with telescope. It can be locked at any position by vertical clamp (2), and fine adjustment may be done by vertical tangent screw. The vertical circle is graduated from 00 to 3600, sometimes it is divided into four quadrants from 00 to 900.

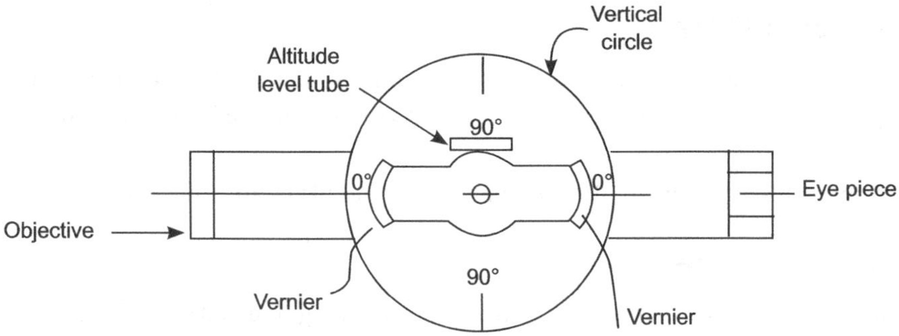

Fig. 2.2: Vertical circle with verniers and telescope

2.3.3 The Index Frame or Vernier Frame

Vernier Frame (21) is also called T frame as it is T-shaped, consisting of a vertical leg called as clipping arm and a horizontal arm called as index arm or vernier arm. Two verniers are provided at both ends of index arm to read the vertical circle. The centre of index arm is coincided by trunnion axis. The index frame remains fixed in front of the vernier circle. When telescope is rotated in vertical plane, the vertical circle also moves relative to verniers facilitates reading of main scale of vertical circle. However, the index arm may be rotated slightly for adjustment purpose by a clip screw fitted on the clipping arm. Glass magnifiers are placed above each vernier to magnify the reading. A bubble tube is fixed on the top of the index frame, it is also called altitude bubble (23). [See Fig. 2.3]

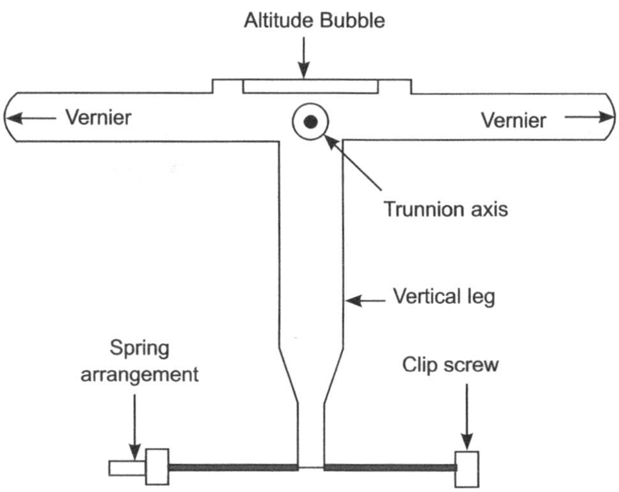

Fig. 2.3: Parts of Vertical Frame

2.3.4 The Standards

Two standards (4) are provided on upper plate. They are A- shaped, hence also called A- frame , they support telescope and allow it to rotate about trunnion axis.

2.3.5 The Upper Plate

This is also known as vernier plate (8). It is attached to inner spindle and carries two verniers with glass magnifier at opposite ends. It carries upper clamp (6) for locking it to the lower plate and a tangent screw for finer adjustment.

When the upper clamp is tightened, both upper and lower plates are attached and moves together.

2.3.6 The Lower Plate

This is also called as scale plate (9) as it carries horizontal circular graduated arc (19) graduated from 00 to 3600. It is attached to the outer spindle and carries lower clamp (10) and a tangent screw. If lower clamp is loosened and upper clamp is tightened, both plates will rotate together.

Similarly, if lower clamp is tightened and upper clamp is loosened then, only upper plate will move and lower plate will be fixed with tribrach plate.

2.3.7 Plate Level

Plate level (20) is carried by the upper plate which is parallel to the trunnion axis. The plate level is used for levelling the instrument.

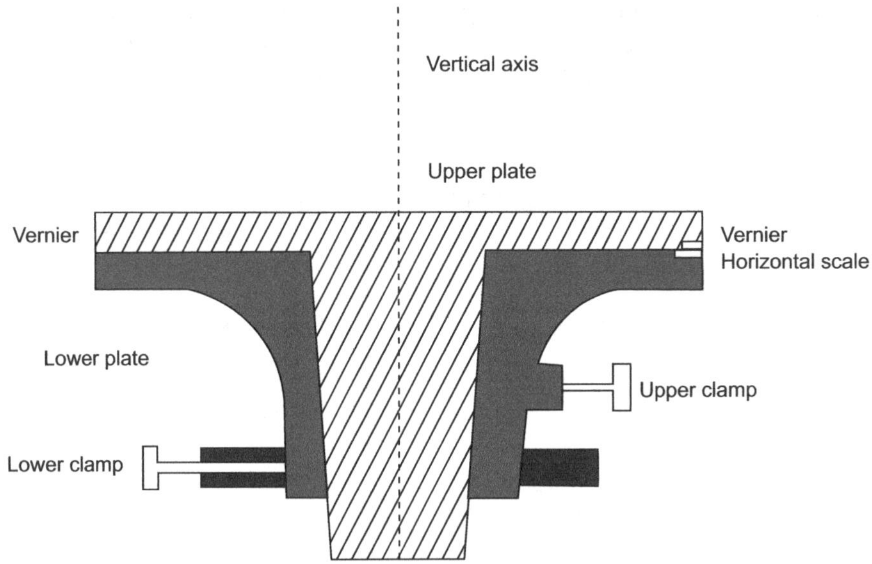

Fig. 2.4: Cross-section along the length through upper and lower plate

2.3.8 The Levelling Head

The levelling head (18) is provided with two parallel triangular plates known as tribrach plates. The upper triangular plate is known as upper tribrach plate (11) and is used to level the instrument with the help of foot screws (12) provided at its three corners. The lower triangular plate is called as lower tribrach plate or foot plate (13) and is attached to the tripod top to support the instrument on the tripod.

Fig. 2.5: Photograph of Transit Theodolite

2.3.9 Tripod

The Theodolite is supported on tripod (14). The tripod head carries an external screw (17), which is screwed to internal screw of foot plate of levelling head.

2.3.10 Plumb Bob

Plumb bob (15) is used to place the Theodolite on the ground station in such a way that vertical axis (16) of the instrument is exactly above the ground point. The bob is suspended from the hook fitted at inner spindle touching the ground. Moving the tripod and adjusting the legs of it , the bottom tip of the bob is brought exactly on ground point. This process is called centering.

2.4 TECHNICAL TERMS USED IN TRANSIT THEODOLITE

(i) **Transiting:** Transiting is also known as reversing or plunging. It is a process of rotating the telescope about its horizontal axis through 180° in the vertical plane.

(ii) **Swinging the telescope**: It is turning the telescope about its vertical axis in the horizontal plane. A swing is called right if the telescope is rotated in clockwise direction and a swing is called left if the telescope is rotated in anti-clockwise direction.

(iii) **Face Left**: If the vertical circle of the instrument is on the left side of the observer while sighting an object, the position is called the face left and the observation taken during this position is called as the face left observation.

(iv) **Face Right:** If the vertical circle of the instrument is on the right side of the observer while sighting an object, the position is called the face right and the observation taken during this position is called as the face right observation.

2.5 FUNDAMENTAL AXES OF TRANSIT THEODOLITE AND THEIR RELATIONSHIP

Fundamental Axes

1. Axis of the Level Tube / Bubble Line: It is a tangent to the longitudinal curve of the level tube at the centre of the tube. It is horizontal when the bubble is in the centre of the bubble tube.
2. Vertical Axis: It is the axis about which the telescope can rotate in the horizontal plane. The upper and lower plates rotate about this axis.
3. Horizontal Axis / Trunnion Axis: It is the axis about which the telescope can rotate in the vertical plane.
4. Line of Sight (LOS)/Line of Collimation (LOC): It is an imaginary line joining the intersection of the cross- hairs of the diaphragm to the optical centre of the objective and its continuation

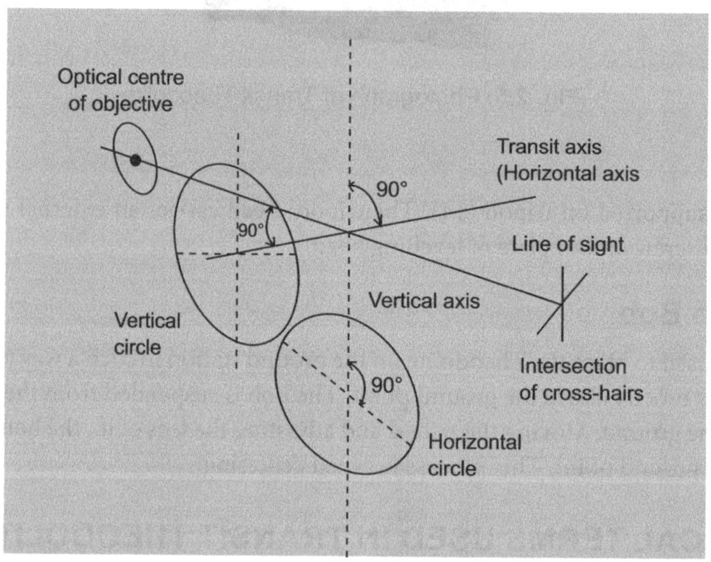

Fig. 2.6: Fundamental Axes of Transit Theodolite and their Relation

Relationship – (see Fig. No. 2.6)

- The axis of the level tube must be perpendicular to the vertical axis.
- The line of collimation must lie in a plane perpendicular to the horizontal axis. Line of collimation, Vertical axis and Horizontal axis must intersect at a point.

- The Horizontal axis must be perpendicular to the Vertical axis.
- The axis of altitude level must lie in a plane parallel to the line of collimation

2.6 READING OF VERNIER OF TRANSIT THEODOLITE

A theodolite is provided with two verniers diagonally opposite to each other (at 180° difference) on the upper plate. For ordinary work, reading of only one is taken, whereas for precise work, reading of both verniers are taken. Later practice minimises errors due to imperfection of subdivisions and eccentricity of circular scale. The main scale is circular, placed on lower plate, graduated from 0° to 360°. Each degree is divided in three equal parts, hence least count of main scale is 20'. The vernier scale is graduated from 0' to 20'. Each minute is divided into three equal parts, hence least count of vernier is 20".

For taking observation, first reading of main scale is taken in degrees and minutes against the zero of the vernier, then vernier reading would be coinciding division of vernier with any main scale division multiplied by the least count of the vernier. The main scale reading is added to vernier scale reading to get the final reading. For example, main scale reading against the zero of vernier is 155° 20' and coinciding division of vernier is 43 then final reading would be 131° 20' + 43 × 20" = 131° 34' 20" (Fig. 2.7)

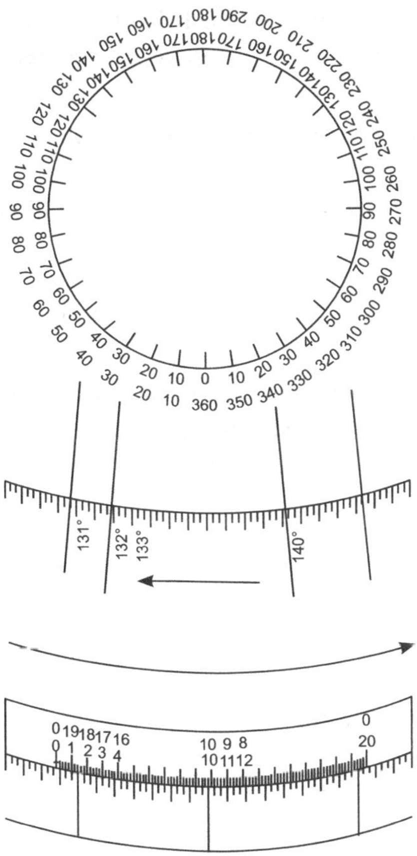

Fig. 2.7: Reading of Vernier of Transit Theodolite

2.7 TEMPORARY ADJUSTMENT OF TRANSIT THEODOLITE

Temporary adjustments of theodolite include:

1. Setting on the station
2. Levelling Up
3. Parallax Elimination

2.7.1 Setting of Theodolite

Firstly, fix the instrument on the tripod and keep the instrument over the station by spreading the legs of the tripod ensuring convenient height of the tripod. Now do the centring of instrument with the help of plumb-bob, by moving the tripod in such a way that the bottom tip of plumb-bob should touch ground point of the station. Sometimes centring is done in windy areas with the help of optical plummet by moving the tripod in such a way that narrow laser beam should exactly fall on ground point of the station. Then approximate levelling is done by adjusting the legs of the tripod.

2.7.2 Levelling of Theodolite

Accurate levelling of the theodolite is done with the help of foot screws and plate level. Two types of levelling screws heads are used by which levelling is done; the process is as under.

Three levelling screws head

1. Place the plate level parallel to a line joining any two levelling screws (say A & B) by turning the upper plate.
2. Now, rotate these two screws (A & B) in opposite directions uniformly until the bubble of the plate level comes in the centre.
3. Rotate the upper plate by 90°.
4. Now, rotate the third levelling screw in such a way, that the bubble of plate level comes in the centre.
5. Repeat steps (1) to (4) till the bubble is in centre in both positions (2) and (4) simultaneously.
6. Now rotate the instrument by 180° if the bubble remains in the centre, then the instrument is called levelled. If not, it needs permanent adjustment.

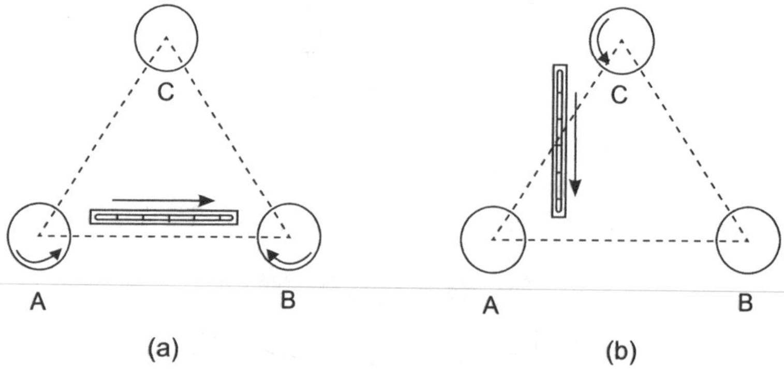

Fig. 2.8: Levelling with three-foot screw

Four levelling screws head

1. Place the plate level parallel to a line joining any two diagonally opposite levelling screws (say B & D) by turning the upper plate.
2. Now, rotate these two screws (B & D) in opposite directions uniformly until the bubble of the plate level comes in the centre.
3. Rotate the upper plate by 90° so that the plate level is parallel to the line joining to other two diagonally opposite screws (say A & C).
4. Now, rotate these two screws (A & C) in opposite directions uniformly until the bubble of the plate level comes in the centre.
5. Repeat steps (1) to (4) till the bubble is in centre in both positions (2) and (4) simultaneously.
6. Now rotate the instrument by 180° if the bubble still remains in the centre of its run, then the instrument is called levelled. If not, it needs permanent adjustment.

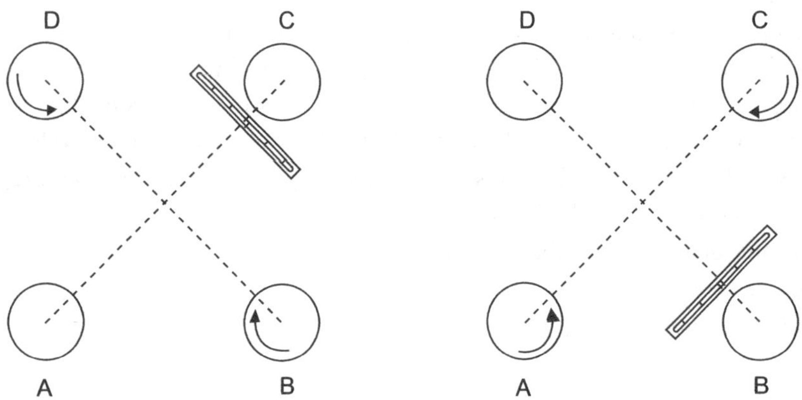

(a) (b)

Fig. 2.9: Levelling with four-foot screw

2.7.3 Elimination of Parallax

The parallax occurs if the image formed by the objective doesn't lie in the plane of the cross hair and the object is not clearly visible through telescope. The parallax can be eliminated by focusing of the eyepiece and the objective.

Focusing of the eyepiece

This can be done by placing the piece of white paper in front of the objective or towards the sky and rotate the eyepiece in such a manner that the cross hairs are visible clearly.

Focusing of the objective

This can be done with the help of a focusing screw. Sight the telescope of theodolite towards the object to be viewed and rotate the focusing screw in such a manner that the object is visible clearly while sighting through the eyepiece. The image so formed is in the plane of cross-hairs.

2.8 MEASUREMENT OF HORIZONTAL ANGLE BY TRANSIT THEODOLITE

2.8.1 Direct Method

The measurement of angle POQ is done as under (Fig. 2.10)

1. Set up the Theodolite on station O and level it
2. Keep the vertical circle at left (face left case)
3. Release the upper clamp and turn upper plate till the zero of any vernier (say A) of upper plate coincides with the zero of main scale of lower plate. Tighten the clamps of both the plates and coincide the two zeros exactly by turning the upper tangent screw. Take the readings of both verniers. The readings of verniers A and B would be 00 and 1800 respectively.
4. Release the lower clamp and sight the station P by swinging the telescope. Tighten the lower clamp and bisect the station point P accurately by lower tangent screw.
5. Release the upper clamp and swing the telescope in clockwise direction and sight the station point Q Tighten the upper clamp and bisect the point Q accurately by upper tangent screw.
6. Take readings of both verniers. The reading of vernier A gives the angle POQ directly while reading of vernier B gives the angle POQ by deducting 1800. Take average of the two values of angle POQ
7. Keep the vertical circle at right by transiting the telescope (face right case) and repeat the above process and determine the angle POQ.
8. Take mean value of angle POQ obtained by steps (6) and (7)

Note: The zero of the vernier is initially set on zero of main scale for convenience purpose only, it can be set at any convenient reading.

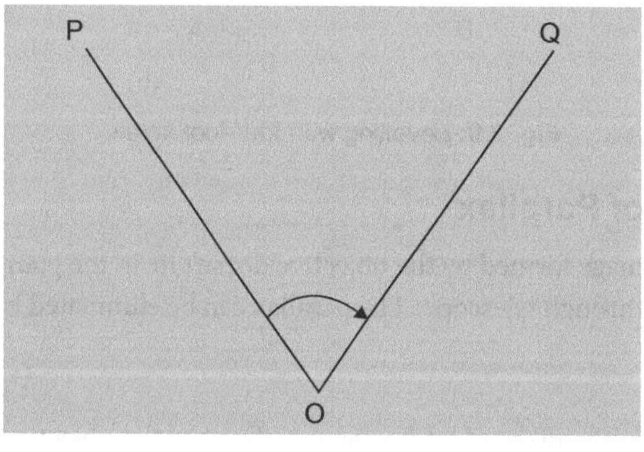

Fig. 2.10: Measurement of horizontal angle

2.8.2 Method of Repetition

To measure a horizontal angle more accurately than that is obtained by the least count of the vernier, the method repetition is used. In this method the angle is measured twice or more times with the vernier to remain clamped at the end of each measurement instead of setting it back at 0° when sighting previous station. Hence the angle reading is multiplied by number of repetitions mechanically. The multiplied reading is divided by number of repetitions to get the mean angle.

The measurement of angle POQ is done as under:

1. Set up the Theodolite on station O and level it
2. Keep the vertical circle at left (face left case)
3. Release the lower clamp and turn upper plate till the zero of any vernier (say A) of upper plate coincides with the zero of main scale of lower plate. Tighten the clamps of both the plates and coincide the two zeros exactly by turning the upper tangent screw. Take the readings of both verniers. The reading of vernier A would be 0°, note the reading of vernier B.
4. Release the lower clamp and sight the station P by swinging the telescope. Tighten the lower clamp and bisect the station point P accurately by lower tangent screw.
5. Release the upper clamp and swing the telescope in clockwise direction and sight the station point Q .Tighten the upper clamp and bisect the point Q accurately by upper tangent screw.
6. Take readings of both verniers. The reading of vernier A gives the angle POQ directly while reading of vernier B gives the angle POQ by deducting 1800. Take average of the two values of angle POQ
7. Release the lower clamp of lower plate and sight the station P again by swinging the telescope in clockwise direction. Tighten the clamp of lower plate and bisect the station point P accurately by tangent screw of lower plate. In this operation, vernier readings will not be changed as there is no relative motion between upper and lower plate.
8. Release the upper clamp and swing the telescope in clockwise direction and sight the station point Q . Tighten the upper clamp and bisect the point Q accurately by upper tangent screw. Take the readings of the verniers. The vernier will read twice the angle POQ.
9. Repeat the process for required number of times, say 03 and average angle for face left case will be equal to final reading divided by 03
10. Keep the vertical circle at right by transiting the telescope (face right case) and repeat the above process 03 more times and determine the average angle for face right case by dividing the final reading by 03
11. Determine average horizontal angle by taking mean of two angles obtained by steps (9) and (10) for face left and face right case.

Errors eliminated by method of repetition

(i) The errors due to eccentric centres of the plates and eccentricity of verniers are rectified by taking both verniers readings.

(ii) The error due to the line of collimation is not being perpendicular to the trunnion axis is eliminated by taking both face readings

(iii) The error due to inaccurate graduation is rectified by reading the angle at different parts of the graduated circle.

(iv) Errors due to inaccurate bisection of the objects are counter balanced to some extent in different observations.

2.9 MEASUREMENT OF VERTICAL ANGLE BY TRANSIT THEODOLITE

A vertical angle is defined as an angle which the inclined line of sight to an object makes with the horizontal. If the object to be sighted is above the horizontal plane, then the angle subtended is called as the angle of elevation and if object is below the horizontal plane, then the angle is called angle of depression.

The measurement of vertical angle is done as under (Fig. 2.11)

1. Set the instruments at the station B and level it with the plate level
2. Release the vertical clamp and rotate the telescope in vertical plane till zero of any vertical vernier coincides with the zero of vertical circle. Tighten the vertical clamp and coincide the two zeros exactly by turning the vertical tangent screw . Check bubble of altitude level, if it is not in centre, bring it in centre with the help of the clip screw.
3. Release the clamp of vertical circle and rotate the telescope in vertical plane to sight the object A. Tighten the vertical clamp and use vertical tangent screw to bisect the object accurately.
4. Read both the vertical verniers C and D. The mean of readings of two verniers will give the value of the required angle
5. Change the face and repeat the process to get another value of required angle
6. Take average of the two values of required angle obtained in steps (4) and (5).

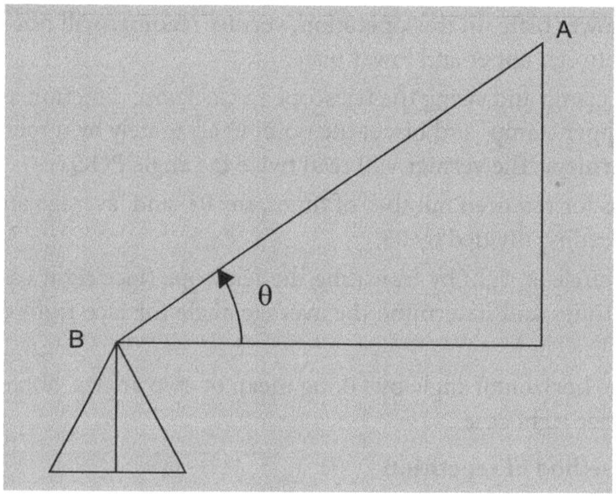

Fig. 2.11: Measurement of vertical angle

2.10 SOME FIELD APPLICATIONS OF TRANSIT THEODOLITE

2.10.1 Measurement of Direct Angle

An angle measured from previous line to next line in clockwise direction is called Direct Angle. The angle ABC is measured as under (Fig. 2.12)

(i) Set the instrument at B and level it. Set the reading of graduated circle to zero against vernier A by turning the upper plate with face left.

(ii) Release the lower clamp and swing the telescope to sight station A. Tighten the lower clamp and bisect A accurately by lower tangent screw

(iii) Release the upper clamp and swing the telescope clockwise to sight the station C. Tighten the upper clamp and bisect station C accurately by upper tangent screw. Take readings of both verniers

(iv) Release the lower clamp and swing the telescope to sight station A again and lock the lower clamp.

(v) Release the upper clamp and swing the telescope clockwise to sight the station C again.

(vi) The angle measured is double of the required angle ABC, hence final reading is divided by two to get more accurate value of angle ABC.

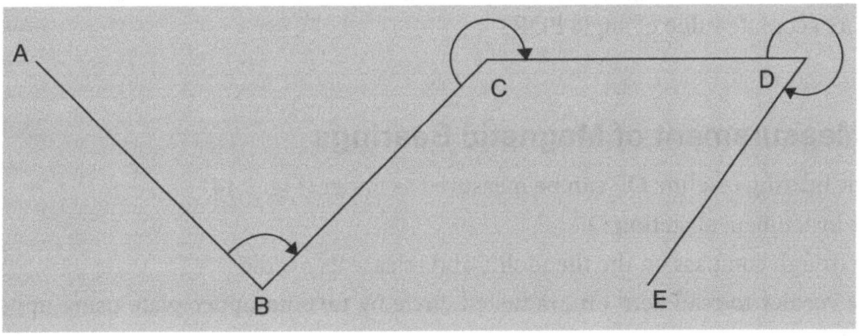

Fig. 2.12: Measurement of Direct Angle

2.10.2 Measurement of Deflection Angle

An angle measured from prolongation of previous line to next line is called Deflection Angle. When the angle measured clockwise, it is called right (R) deflection angle whereas measured anti-clockwise, it is called left(L) deflection angle. The value of deflection angle varies from 0° to 180°. In Fig. 2.13, the deflection angle is α(R) at Q and β(L) at R. The deflection angle is measured as under.

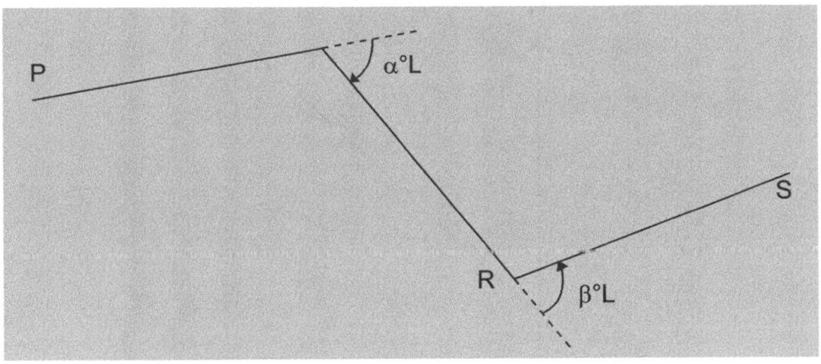

Fig. 2.13: Measurement of Deflection Angle

(i) Set the theodolite at station Q and level it.

(ii) Set the reading of graduated scale to zero against vernier A by turning upper plate and with the help of upper clamp and tangent screw.

(iii) Take back sight at station P ensuring vernier A reads zero

(iv) Transit the telescope keeping linc of sight is in the direction of PQ produced keeping vernier A reads zero.

(v) Release the upper clamp and rotate the telescope clockwise to sight station R with finer adjustment by tangent screw.

(vii) Take readings of both verniers

(viii) Release the lower clamp and turn the telescope to sight P again keeping the verniers to read same readings and lock lower clamp.

(ix) Transit the telescope and release the upper clamp and rotate the telescope to sight station R. Take readings of vernier.

(x) The angle measured is double of the required angle PQR, hence final reading is divided by two to get more accurate value of angle PQR.

2.10.3 Measurement of Magnetic Bearings

The magnetic bearing of a line OP can be measured as under (Fig. 2.14)

(i) Set the instrument at station O

(ii) Fit the trough compass on the theodolite and release its needle

(iii) Set the vernier to read zero on graduated circle by turning upper plate using upper clamp and tangent screw

(iv) Release the lower clamp and rotate the lower plate till magnetic needle rest in N-S direction. Lock the lower clamp and bring the needle in true magnetic north by lower tangent screw.

(v) Release the upper clamp and bring the telescope in the line of station P. Lock the upper clamp and bisect station P accurately using upper tangent screw. Take readings of both verniers

(vi) Change the face of the instrument and repeat the steps from (ii) to (v) The average of two values obtained from readings of two faces is correct value of the bearing of line OP.

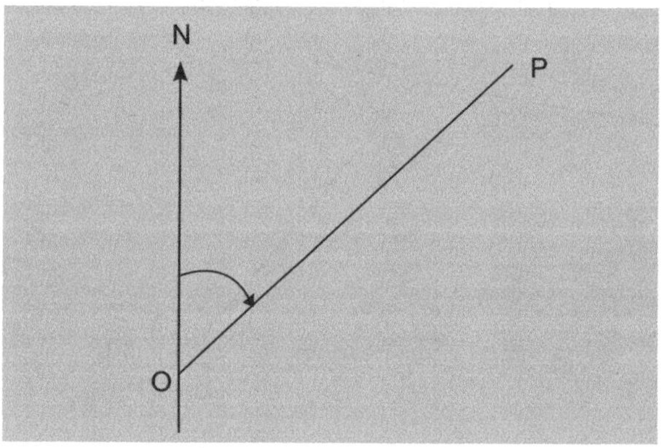

Fig. 2.14: Measurement of Magnetic Bearings

2.10.4 Prolonging a Straight Line

The prolonging the line AB can be done as under (Fig. 2.15)

(i) Set the theodolite at A and level it

(ii) Sight B accurately by turning lower plate and using lower clamp and tangent screw

(iii) Looking from telescope, direct the surveyor to bring the ranging rod exactly at C such that B and C in a line.

(iv) Shift the instrument at B and repeat above steps

(v) Continue the process until last point E is established.

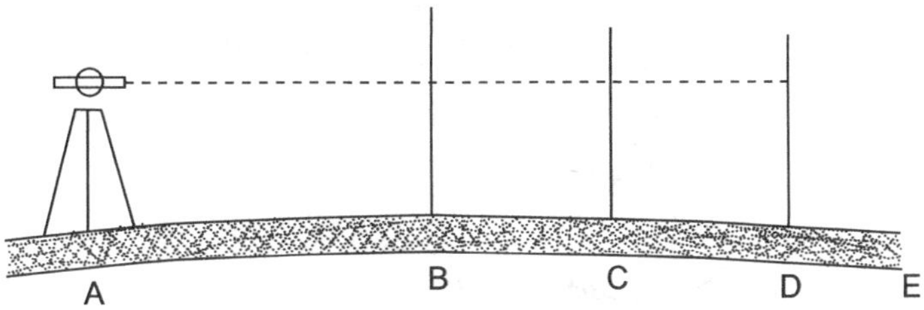

Fig. 2.15: Prolonging a Straight Line

2.11 THEODOLITE TRAVERSING

Traversing is a type of surveying which consists of number of connected survey lines. The direction of the lines is measured with the help of chain, compass and theodolite whereas length is measured by tape or chain. A traverse is said to be closed if it returned to starting point whereas it is called open traverse if it does not return to starting point and ends elsewhere. The closed traverse is suitable to locate the boundaries of ponds, lakes, forests etc. whereas open traverse is useful for locating long narrow strips of land for roads, canals and coastal lines.

When theodolite is used for direction or angle measurement, it is called theodolite traversing. Theodolite traversing is done by following methods.

2.11.1 Traversing by Included Angles

In this method, the angle between two survey lines is measured directly by taking backsight at preceding station. The greater accuracy is obtained by taking readings of both verniers, taking both face observations, and using method of repetition. It is preferred to measure the angles clockwise as the graduations on graduated circle are marked in clockwise direction. The angle measured clockwise from previous station may be interior angle or exterior angle depending on direction of progress of survey. If direction of progress of work is anticlockwise, angle measured clockwise will be interior angle (Fig. 2.16) and if direction of progress of work is clockwise, angle measured clockwise will be exterior angle (Fig. 2.17).

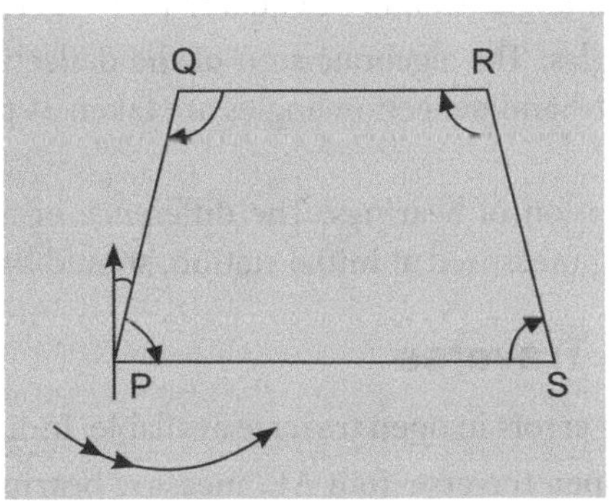

Fig. 2.16: Included angle method -Interior angle measured clockwise

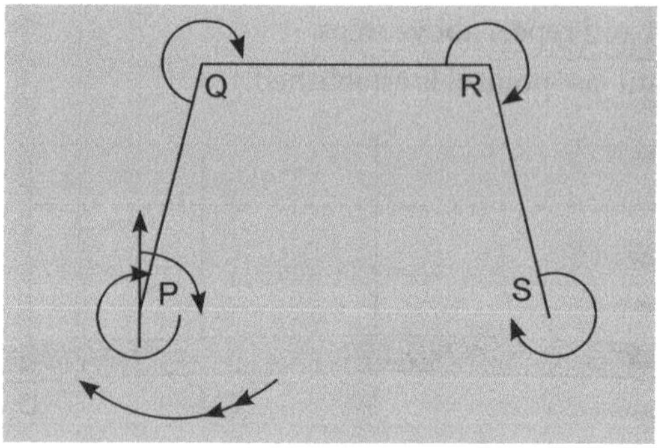

Fig. 2.17: Included angle method - Exterior angle measured clockwise

2.11.2 Traversing by Deflection Angles

The deflection angle and method of its measurement is described in section (). The traversing by this method is suitable where survey lines make small deflection angles like roads, railways and pipelines etc. The attention must be given in recording and plotting of correct deflection angle whether it is right or left deflection angle. It is preferred to read included angles clockwise from back station except in specialised work where deflection angles are required to read. The lengths of lines are measured by high quality steel tape.

2.12 CHECKS IN TRAVERSING

2.12.1 Checks in Closed Traverse

The errors occur in closed traverse may be two types:

(i) Linear errors

(ii) Angular errors

(i) Linear errors: Linear errors may be checked by chaining each survey line twice, preferably in reverse direction second time on different dates and by different surveyors.

(ii) Angular errors: Angular errors may be checked as under:

(a) Traversing by included angles: The sum of interiors angles of traverse should be equal to (2N – 4) right angles and sum exterior angles should be equal to (2N + 4) right angles.

(b) Traversing by deflection angles: The algebraic sum of the deflection angles of traverse should be equal to 360° provided right-hand deflection angles are taken as positive and left-hand deflection angles are taken as negative

(c) Traversing by direct observation of bearings: The difference of fore bearing of the last line of a traverse and its back bearing measured at initial station, should be equal to 180°.

2.12.2 Checks in Open Traverse

There is no direct check of angular errors in open traverse available. Indirect checks can be used as under

(a) In Fig. 2.18, ABCDE is an open traverse. Join AE, measure bearing of AE at A, similarly measure bearing of EA at E, the difference of two bearings should be 180°.

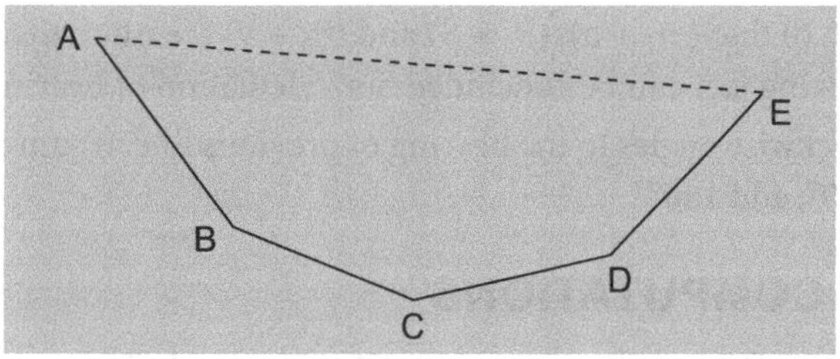

Fig. 2.18: Checks for open traverse - Case 1

(b) In Fig. 2.19 ABCD is an open traverse. Take any point P as shown in the fig., read the bearings of P from consecutive stations A, B, C and D and draw lines AP, BP, CP and DP by measured bearings and as a check, these lines should pass through one point.

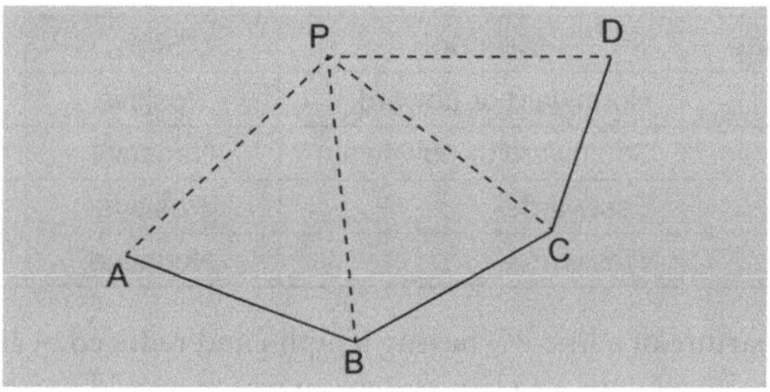

Fig. 2.19: Checks for open traverse - Case 2

2.13 CALCULATIONS OF BEARING FROM ANGLES

If in a traverse included angles between successive lines are known, the bearings of the lines may be calculated if bearing of any one line is also known.

In an open traverse ABCDEF, the bearing of line AB is θ_1 and included angles between lines measured clockwise are α, β, γ and δ then

The bearing of the next line BC = $\theta_2 = \theta_1 + \alpha - 1800$

The bearing of the next line CD = $\theta_3 = \theta_2 + \beta - 1800$

The bearing of the next line DE = $\theta_4 = \theta_3 + \gamma - 1800$

The bearing of the next line EF = $\theta_5 = \theta_4 + \delta + 1800$

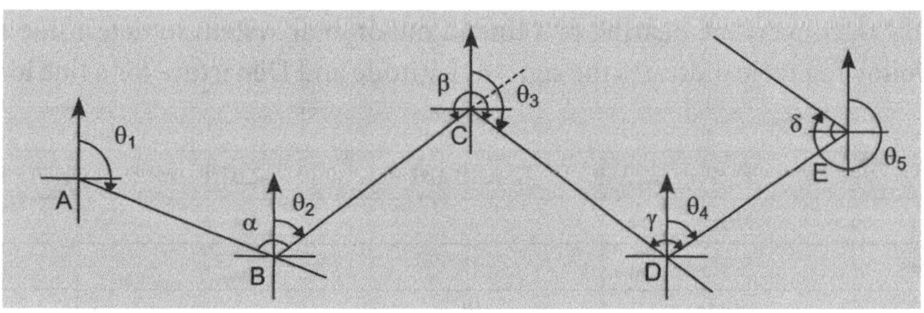

Fig. 2.20: Calculations of Bearing from Angles

It is clear in the Fig. 2.20 that ($\theta_1 + \alpha$), ($\theta_2 + \beta$) and ($\theta_3 + \gamma$) are more than 180° whereas ($\theta_4 + \delta$) is less than 180°, hence following fact can be concluded for calculation of bearing of next line.

"Add the included clockwise angles to the bearing of previous line, if sum is more than 180°, deduct 180° if sum is less than 180°, add 180°".

2.14 TRAVERSE COMPUTATIONS

2.14.1 Consecutive Co-ordinates: Latitude and Departure

The latitude (L) of a survey line is defined as co-ordinate length of the line measured parallel to a assumed meridian direction. The assumed meridian may be true north or magnetic north or any reference direction. The departure (D) of a survey line is defined as co-ordinate length of the line measured perpendicular to an assumed meridian direction. The following sign convention is adopted.

Consecutive Co-ordinate	Direction	Sign	Termed as
Latitude	Northward or upward	Positive	Northing
Latitude	Southward or downward	Negative	Southing
Departure	Eastward	Positive	Easting
Departure	Westward	Negative	Westing

The Latitude and Departure of a line PQ having length l and reduced bearing θ are given as

$$L = +l\cos\theta \quad \text{and} \quad D = +l\sin\theta \text{ (Fig. 2.21)}$$

They are also called Consecutive Co-ordinates

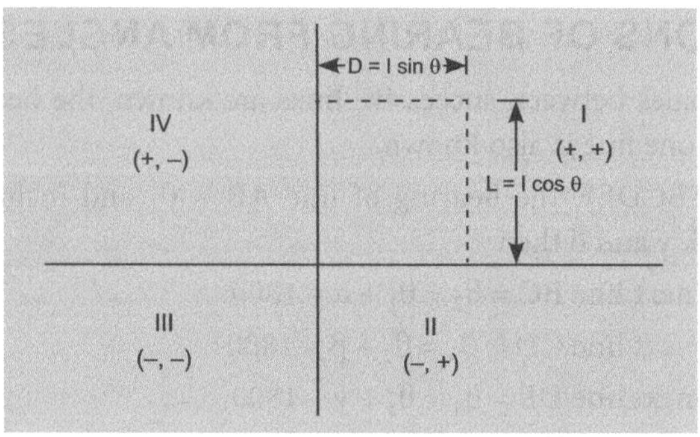

Fig. 2.21: Latitude and Departure

It is necessary to convert the bearing of a line in quadrantal system to determine its Latitude and Departure. The following table indicates the signs of Latitude and Departure for a line located in various quadrants.

W.C.B	R.B.	Quadrant	Latitude	Departure
0° to 90°	NθE	I	+	+
90° to 180°	SθE	II	-	+
180° to 270°	SθW	III	-	-
270° to 360°	NθW	IV	+	-

2.14.2 Independent Co-ordinates

The total latitude and departure of a point with respect to a common origin are called as independent co-ordinates or total co-ordinates of that point. A set of vertical and horizontal axes passing through a traverse station located at westernmost end is taken as reference axes and western most station is considered as common origin.

The independent co-ordinates of a point may be determined by algebraic sum of all the latitudes and the departures of the lines between that point and the origin.

2.14.3 Closing Error in Traverse

Due to the errors in the measurements of angles and distances in the field, end point and starting point does not coincide to each other in closed traverse. This error is called as closing error (Fig. 2.22). In perfect closed traverse, algebraic sum of latitude and algebraic sum of departure is zero i.e. $\Sigma L = 0$, $\Sigma D = 0$. The closing error is found by determining ΣD and ΣL and thus closing error $AA' = e = \sqrt{[(\Sigma L)^2 + (\Sigma D)^2]}$.

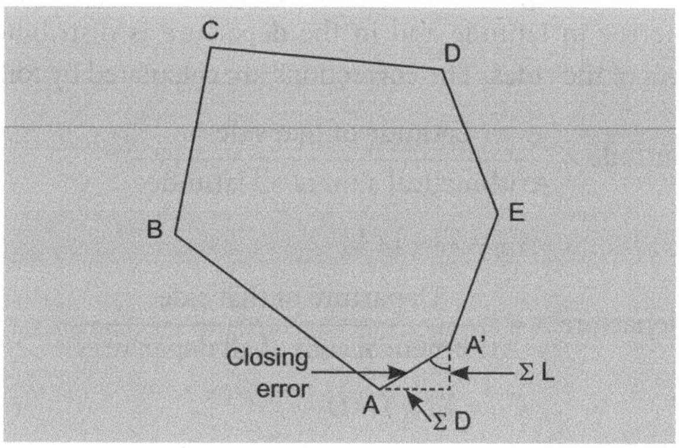

Fig. 2.22: Closing Error in Traverse

2.15 BALANCING OF TRAVERSE

The balancing of a close traverse consists of applying corrections to latitudes and departures so that total latitude and total departure are zero i.e. $\Sigma D = 0$, $\Sigma L = 0$. This is applicable only for close traverse.

The common methods of balancing of a traverse are as under:

(i) Bowditch method

(ii) Transit method

(iii) Graphical method

(iv) Axis method

Here, first two methods are discussed:

2.15.1 Bowditch method

This method is based on Bowditch rule which assumes that the linear errors are proportional to $\sqrt{l}$ and angular errors are inversely proportional to $\sqrt{l}$ where l is the length of a line.

The Bowditch rule is also known as compass rule. This method is generally used when angular and linear measurements are taken with equal precision. The total error in latitude and in the departure is distributed in proportional to the length of the sides. The corrections are calculated by following equations

$$\text{Correction in latitude of a line} = \text{Total error in latitude} \times \frac{\text{Length of the side}}{\text{Perimeter of traverse}}$$

$$C_L = \Sigma L\ X_l/\Sigma_l$$

$$\text{Correction in Departure of a line} = \text{Total error in Departure} \times \frac{\text{Length of the side}}{\text{Perimeter of traverse}}$$

$$C_D = \Sigma D\ X_l/\Sigma_l$$

2.15.2 Transit Method

The transit rule is used when angular measurements are more precise in comparison of linear measurements. The total error in latitude and in the departure is distributed in proportional to the latitudes and the departures of the sides. The corrections are calculated by following equations:

$$= \text{Total error in latitude} \times \frac{\text{Latitude of that side}}{\text{Arithmetical sum of all latitudes}}$$

$$C_L = \Sigma L_X\ L/\ L_T$$

$$= \text{Total error in departure} \times \frac{\text{Departure of that side}}{\text{Arithmetical sum of all departures}}$$

$$C_D = \Sigma D_X\ D/\ D_T$$

2.16 GALES TRAVERSE TABLE

Traverse computations are mostly done in the form of Gales Traverse Table. The following steps are adopted to complete the traverse computations.

(i) Adjust the included interiors angles in such a way that the sum of interiors angles of traverse should be equal to (2N – 4) right angles and sum exterior angles should be equal to (2N + 4) right angles.

(ii) Read the bearing of one line and calculate the bearing of other lines using included angles and convert the bearing in quadrantal form.

(iii) Calculates latitudes and departures of survey lines (consecutive coordinates).

(iv) Calculate ΣL and sigma ΣD.

(v) Apply necessary corrections by applying Bowditch rule or transit rule so that ΣL = 0 and ΣD = 0.

(vi) Using corrected consecutive co-ordinates, calculate the independent co-ordinates of the points. The origin is so selected that entire traverse lies in N-E quadrant. This is done for plotting of the traverse on a sheet with the left-hand bottom corner of the sheet as the origin.

Table 2.1 illustrates the gales traverse table completely

Table 2.1: Gales Traverse Table

			Consecutive Coordinate				Corrections				Corrected Consecutive Coordinate					Independent Coordinate	
Line	Length (m)	RB	Latitude		Departure		Latitude		Departure		Latitude		Departure		Point	Latitude	Departure
			N	S	E	W	N	S	E	W	N	S	E	W		N	E
															A	1000	1000
AB	250	S50°E	–	160.7	191.51	–	–	–0.013	–0.008	–	–	160.713	191.518	–	B	839.312	1191.502
DC	600	N42°E	445.88	–	401.48	–	0.003	–	–0.019	–	445.811		401.499	–	C	1285.182	1592.96
CD	100	N43°E	73.14	–	–	68.2	0.005	–	–	0.003	73.135		–	68.197	D	1358.317	1524.766
DA	635.46	S55°40E	–	358.4	–	524.74	–	0.0032	–	0.002	–	358.403	–	524.738	A	989.914	1000.028
	Σ		519.02	519.1	592.99	592.94	0.008	–0.0162	–0.027	0.005	519.012	519.116	593.017	592.935			

Note:

1. The correction of latitude and departure are calculated by Bowditch rule.
2. The latitude and departure of origin A are assumed as (1000, 1000)

UNIT SUMMARY

- Theodolite is used for – To measure the vertical & Horizontal angles.
- Theodolites may be classified two ways
  - (A) Classification based on movement of telescope:
    - (i) Transit Theodolite
    - (ii) Non-transit Theodolite
  - (B) Classification based on arrangement of reading observations
    - (i) Vernier Theodolite
    - (ii) Micrometer Theodolite
    - (iii) Optical Theodolite
    - (iv) Electronic or Digital Theodolite
- Measurement of horizonal angle is done by direct and repetition method
- Repetition method removes the errors due to eccentric centres of the plates, inaccurate line of collimation, inaccurate graduation and inaccurate bisection.
- Theodolite traversing - Traversing is a type of surveying which consists of number of connected survey lines. The angles are measured by Theodolite in Theodolite traversing
- A traverse is said to be closed if it returned to starting point whereas it is called open traverse if it does not return to starting point and ends elsewhere.
- Traversing is done by included angle method and deflection angle method. The included angle method is preferred over the deflection angle method
- Latitude = L cosθ, Departure = L sinθ
- Transit rule for adjusting the closing error
  - (i) Correction to Latitude of any side

$$= \text{Total error in latitude} \times \frac{\text{Latitude of that side}}{\text{Arithmetical sum of all latitudes}}$$

  - (ii) Correction to departure of any side

$$= \text{Total error in departure} \times \frac{\text{Departure of that side}}{\text{Arithmetical sum of all departures}}$$

- Bowditch rule for adjusting the closing error

$$\text{Correction in latitude of a line} = \text{Total error in latitude} \times \frac{\text{Length of the side}}{\text{Perimeter of traverse}}$$

$$\text{Correction in Departure of a line} = \text{Total error in Departure} \times \frac{\text{Length of the side}}{\text{Perimeter of traverse}}$$

- Traverse computations are usually done in tabular form by Gale's traverse table.

EXERCISE

Multiple Choice Questions

1. How many levelling screws available in theodolite?
   (a) 2 or 5
   (b) 3 or 4
   (c) 4 or 5
   (d) 5 or 3
2. In the transit theodolite, telescope can be revolved
   (a) 90 degrees
   (b) 180 degrees
   (c) 270 degrees
   (d) 360 degrees
3. How many spirit levels are available in theodolite?
   (a) 1
   (b) 2
   (c) 3
   (d) 4
4. What is the example of natural error in plan traverse survey?
   (a) Improper levelling
   (b) Improper reading
   (c) Wind effect
   (d) Improper recording
5. What is not the example of personal error?
   (a) Wrong recording
   (b) Wrong reading
   (c) Improper levelling
   (d) Eccentricity of Vernier
6. The least count of Vernier theodolite
   (a) 10 second
   (b) 20 second
   (c) 25 second
   (d) 30 second
7. ________ is the instrument designed for measurement of horizontal and vertical angels.
   (a) Dumpy level
   (b) Theodolite
   (c) Compass
   (d) Chain
8. If one Vernier reading is 0 degrees, then another Vernier will show ________ degree.
   (a) 90 degrees
   (b) 0 degrees
   (c) 180 degrees
   (d) 360 degrees
9. How many types of errors occurs in theodolite survey?
   (a) 1
   (b) 2
   (c) 3
   (d) 4
10. Inaccurate levelling is the example of ________ Error.
    (a) Instrumental
    (b) Natural
    (c) Personal
    (d) Instrumental and natural
11. Departure can be expressed as
    (a) $D = l \sin\theta$
    (b) $D = l \cos\theta$
    (c) $D = l \tan\theta$
    (d) $D = \cot\theta$

12. Included angels are measured
    (a) Clockwise
    (b) Anticlockwise
    (c) Clockwise and Anticlockwise
    (d) Clockwise or Anticlockwise
13. The sum of the interior angels is equal to ________
    (a) (2n – 4) * 90°
    (b) (2n – 4) * 180°
    (c) (2n + 4) * 90°
    (d) (2n + 4) * 180°
14. What is the height of a building if vertical angle is measured 20 degree 15 minutes 35 seconds and horizontal distance is 50 m where H.I. is 1 m.
    (a) 18 m
    (b) 19.45 m
    (c) 20 .16 m
    (d) 22.35 m
15. Bowditch rule is proposed for
    (a) A closed traverse for adjustment of closing error
    (b) An open traverse for graphical adjustment
    (c) Determination the effect of local attraction
    (d) An open traverse for personal error
16. Which method can also be known as Compass rule?
    (a) Transit method
    (b) Bowditch's method
    (c) Fast needle method
    (d) Direct Method
17. Independent co-ordinates can be calculated by ________.
    (a) Geometric co-ordinates
    (b) Cylindrical co-ordinates
    (c) Spherical co-ordinates
    (d) Consecutive co-ordinates
18. The process of turning the telescope about the horizontal axis in vertical plane is known as
    (a) Transiting
    (b) Reversing
    (c) Plunging
    (d) Swinging
19. The error due to eccentricity of inner and outer axes can be eliminated by
    (a) Taking the mean of reading of both Verniers.
    (b) Taking the mean of reading of both face observations.
    (c) Double sighting
    (d) Taking the mean of several readings distributed over different portions of the graduated circle.
20. Following are the sequence of steps for gales traverse table
    (a) Balancing of traverse by Bowditch transit or third rule and correcting consecutive coordinates
    (b) Computing bearings for included angles and converting WCB to QB
    (c) Computing independent coordinates
    (d) Computing latitude and departure and apply usual check for consecutive coordinates.

ANSWERS													
1.	(b)	2.	(b)	3.	(b)	4.	(c)	5.	(d)	6.	(b)	7.	(b)
8.	(c)	9.	(c)	10.	(c)	11.	(a)	12.	(d)	13.	(a)	14.	(b)
15.	(a)	16.	(b)	17.	(d)	18.	(a)	19.	(d)	20.	(a)		

Short and Long Answer Type Questions

Short Questions:

1. Define Swinging, Transiting, Face left, Face right in context to Theodolite.
2. Write the types and uses of Theodolite.
3. List out the Components of transit Theodolite.
4. Write the short note on the temporary adjustment of transit Theodolite.
5. Write steps of determining vertical angle by Theodolite.
6. Write down errors eliminated by method of repetition.
7. How prolonging of a line is done by theodolite.
8. Explain Latitude & Departure.
9. Write down steps of determining magnetic bearing of a line by Theodolite.
10. Write notes on Bowditch's rule and Transit rule of balancing traverse.
11. Discuss Gale's traverse table computation.

Long Questions:

1. Discuss the components of theodolite & their functions.
2. Draw line diagram of Theodolite and level its parts.
3. Explain Fundamental axes of transit theodolite and their relationship with neat diagram.
4. Discuss the Direct and Repetition method of determining horizontal angle by Theodolite.
5. Discuss the Included angle method and Deflection angle method of traversing.
6. Discuss the checks for closed and open traverse.
7. Discuss traverse computation and balancing of traverse.

Solved Numerical Problems

Numerical 1: Find the interior angles from the bearing of line measured during a theodolite survey of a closed traverse ABCDA.

Line	Fore Bearing
AB	50°
BC	80°
CD	220°
DA	280°

Solution: Interior angle = Bearing of next line – Bearing of preceding line

$\angle A$ = Bearing of line AB – Bearing of line AD

= 50° – (280° – 180°)

= 50° (Anticlockwise)

$\angle B$ = Bearing of line BC – Bearing of line BA

= 80° – (50° + 180°)

= 150° (Anticlockwise)

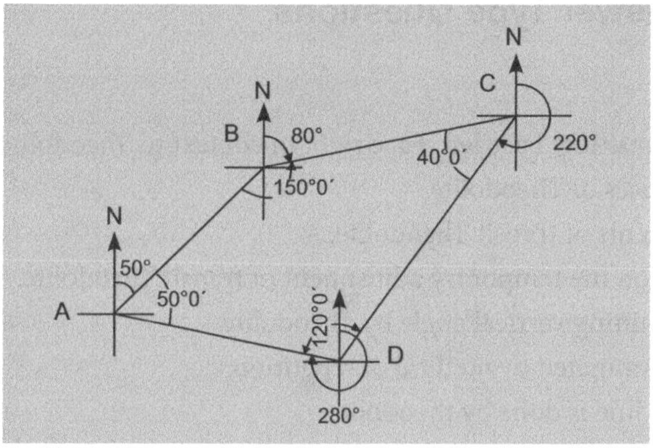

Fig. 2.23: Determination of interior angles

$\angle C$ = Bearing of line CD – Bearing of line CB

= 220° – (80° + 180°)

= 40° (Anticlockwise)

$\angle D$ = Bearing of line DA – Bearing of line DC

= 280° – (220° – 180°)

= 240° (clockwise) (exterior angle)

Hence,

Interior angle = 360° – 240° = 120° (Anticlockwise)

Sum of all angles = $\angle A + \angle B + \angle C + \angle D$ = 50° + 150° + 40° + 120° = 360°

Sum of all angles = (2n-4) × 90°

= 360° = (2 × 4 – 4) × 90°

= 360° = 360° **Hence, ok**

Numerical 2: A closed traverse was conducted by theodolite survey around a building and the following observations were noted. Find the length of survey line DE and line EA.

Survey Line	Length (m)	Bearing
AB	586	97° 30'
BC	462	29° 30'
CD	674	297° 30'
DE	?	230° 30'
EA	?	158° 20'

Solution:

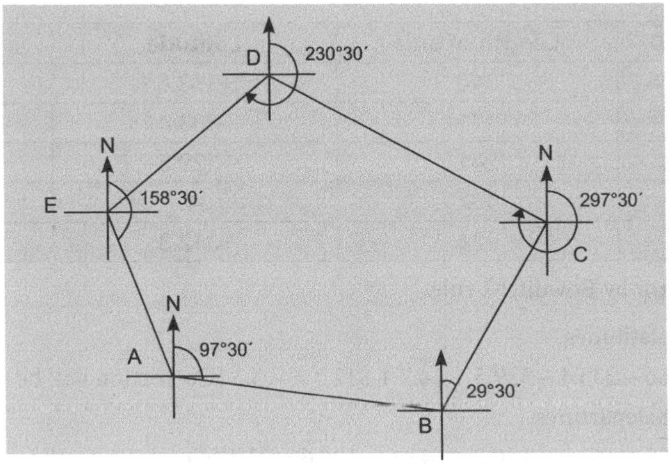

Fig. 2.24: Determination of length of survey lines.

The latitude is expressed as-

$$L = l \cos \theta$$

The Departure is expressed as

$$D = l \sin \theta$$

Survey Line	Length (m)	W.C.B.	R.B.	Latitude	Departure
AB	586	97° 30'	S82° 30'E	–76.48	580.98
BC	462	29° 30'	N29° 30'E	402.10	227.49
CD	674	297° 30'	N62° 30'E	311.21	–597.84
DE	l_1	230° 30'	S50° 30'W	$-0.636\ l_1$	$-0.777\ l_1$
EA	l_2	158° 18'	S21° 42'E	$-0.929 l_2$	$0.369\ l_2$

$\Sigma L = 0$

$= -76.48 + 402.10 + 311.21 - 0.636\ l_1 - 0.929\ l_2 = 0$

$= 0.636\ l1 + 0.929\ l2 = 636.83$ ...(i)

$\Sigma D = 0$

$= 580.98 + 227.49 - 597.84 - 0.771\ l_1 + 0.370\ l_2 = 0$

$= 0.771\ l1 - 0.369\ l2 = 210.63$ ...(ii)

By solving equation (i) & (ii)

$l_1 = 452.88$ m

$l_2 = 375.45$ m

Numerical 3: The abstract sheet of closed traverse ABCDEA are given below-

Side	Length of side	Latitude	Departure
AB	340	+162.6	+422.8
BC	171	–233.4	+163.5
CD	241	–329.5	–130.2
DE	437	+84.7	–541.6
EA	224	+312.3	+106.6

Adjust the closing error by Bowditch's rule.

Solution: Sum of the latitudes,

ΣL = 162.6 – 233.4 – 329.5 + 84.7 + 312.3 = – 3.3 (Correction will be +Ve)

Sum of the Departures,

ΣD = 422.8 + 163.5 – 130.2 – 541.6 + 106.6 = 21.1 (Correction will be –Ve)

Corrections by Bowditch rule:

Correction in latitude of a line = Total error in latitude × (Length/ Perimeter of traverse)

Similarly, Correction in departure in line = Total error in departure × (Length/ Perimeter of traverse)

Perimeter = 340 + 171 + 241 + 437 + 224 = 1413

Calculation of correction to latitudes:

For line AB = 3.3 × (340/1413) = 0.794

For line BC = 3.3 × (171/1413) = 0.40

For line CD = 3.3 × (241/1413) = 0.56

For line DE = 3.3 × (437/1413) = 1.02

For line EA = 3.3 × (224/1413) = 0.52

Calculation of correction to departure:

For line AB = 21.1 × (340/1413) = 5.08

For line BC = 21.1 × (171/1413) = 2.55

For line CD = 21.1 × (241/1413) = 3.60

For line DE = 21.1 × (437/1413) = 6.53

For line EA = 21.1 × (224/1413) = 3.34

Side	Length of side	Latitude	Departure	Correction in Latitude	Correction in Departure
AB	340	+162.6	+422.8	0.794	–5.08
BC	171	–233.4	+163.5	0.40	–2.55
CD	241	–329.5	–130.2	0.56	–3.60
DE	437	+84.7	–541.6	1.02	–6.53
EA	224	+312.3	+106.6	0.50	–3.34

Side	Length of side	Corrected Latitude	Corrected Departure
AB	340	163.40	417.72
BC	171	–233	160.95
CD	241	–328.94	–133.8
DE	437	85.72	–548.13
EA	224	312.8	103.26

Numerical 4 – The abstract sheet of closed traverse ABCDEA are given below:

Side	Length (m)	Bearing
AB	235.40	35°15'
BC	295.50	150°45'
CD	187.40	227°15'

Calculate the length & bearing of line DA.

Solution:

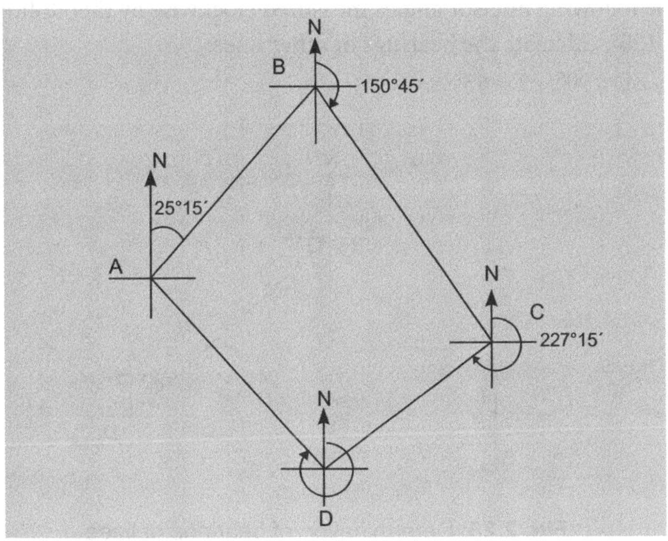

Fig. 2.25: Determination of lengths & bearings of lines

The latitude and departure of any line is given by:

Latitude (L) = l cos θ

Departure (D) = l sin θ

Side	Length (m)	R.B.	Latitude	Departure
AB	235.40	N35°15'E	+192.23	+135.85
BC	295.50	S29°15'E	–257.82	+144.38
CD	187.40	S47°15'E	–127.21	–137.61
SUM			ΣL = –192.8	ΣD = 142.62

Sum of Departure, $\Sigma D = 135.85 + 144.38 - 137.61 = 142.62$

Let, L is latitude and D is Departure of line DA

$$-192.8 + L = 0$$
$$L = 192.8$$
$$142.62 + D = 0$$
$$D = -142.62$$

Length of DA $= \sqrt{L^2 + D^2}$

$$= \sqrt{(192.8)^2 + (-142.62)^2}$$
$$= 239.81 \text{ m}$$ **(Ans)**

$$= \tan\theta = \Sigma D/\Sigma L$$
$$= 142.62/192.8 = 0.739$$
$$= 36°27'51''$$

There is latitude of DA is positive and Departure is negative, the quadrant is N-W.

Bearing of DA = N 36°27'51" W **(Ans)**

Numerical No. 5: The following interior angles measured clockwise by theodolite in a open traverse the bearing of line AB is 120°, calculate the bearings of other lines.

$\angle B = 95°$, $\angle C = 280°$, $\angle D = 90°$, $\angle E = 85°$

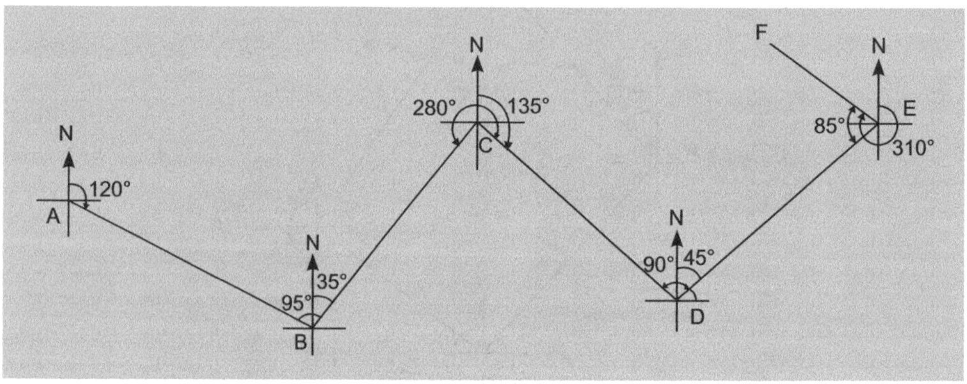

Fig. 2.26: Determination of bearings of lines

Bearing to line AB $= \theta_1 = 120°$

Bearing of line BC $= \theta_2 = \theta_1 + \alpha + 180°$

$$\theta_2 = 120° + 95° - 180°$$
$$\theta_2 = 120° - 85°$$
$$\theta_2 = 35°$$

Bearing of line CD $= \theta_3 = \theta_2 + \beta + 180$

$$\theta_3 = 35 + 280° - 180°$$
$$\theta_3 = 35° + 100°$$
$$\theta_3 = 135°$$

Bearing of line DE = $\theta_4 = \theta_3 + \gamma + 180°$

$\theta_4 = 135° + 90° - 180°$

$\theta_4 = 135° - 90°$

$\theta_4 = 45°$

Bearing of line EF = $\theta_5 = \theta_4 + \delta + 180°$

$\theta_5 = 45° + 85° + 180°$

$\theta_5 = 310°$

Unsolved Numerical Questions

Numerical 1: Find the interior angles of closed transverse ABCDA. The length of lines and W.C.B of the lines are given below.

Line	Length in (m)	W.C.B.
AB	275	46°20'
BC	384	129°30'
CD	367.35	249°44'
DA	236	320°20'

Answer: 96°50' , 59°46' , 109°24' , 94°0'

Numerical 2: A closed traverse was conducted by theodolite survey around a building and the following observations were noted. Find the length of survey line PQ & QM.

Line	Length	Bearing
MN	877	0°0'
NO	721	70°15'
OP	2000	158°35'
PQ	l_1	270°0'
QM	l_2	304°53'

Answer: $l_1 = 1140$, $l_2 = 966.89$

Numerical 3: The abstract sheet of closed traverse ABCDEFA are given below

Line	Length	Bearing
AB	316	52°45'
BC	422.21	68°9'
CD	456.5	110°50'
DE	520.5	246°35'
EF	320.04	229°52'
FA	415.37	303°6'

Adjust the closing error by Transit rule.

Answer:

Line	Corrected Latitude	Corrected Departure
AB	191.32	251.56
BC	157.18	391.909
CD	-162.29	426.694
DE	-206.83	-477.583
EF	-206.27	-244.643
FA	226.888	-347.937
Σ	0	0

Numerical 4: The abstract sheet of closed traverse LMNOPL are given below

Line	Length	Bearing
LM	316.33	52°45'
MN	440.40	84°15'
NO	520.50	219°20'
OP	415.11	221°0'
PL	488.23	349°39'

Adjust the closing error by Bowditch's rule.

Answer:

Line	Corrected Latitude	Corrected Departure
LM	191.475	251.786
MN	44.088	438.174
NO	-402.581	-329.907
OP	-313.273	-272.335
PL	480.288	-87.716
Σ	0	0

PRACTICALS

Experiment No. - 1

Objective: Use transit theodolite to measure Horizontal and Vertical angle by Direct Method.

Required Accessories:

1. Transit Theodolite
2. Tripod stand
3. Plumb- bob
4. Tape for distance measurement
5. Ranging rod

6. Pegs
7. Notebook, scale, pencil etc.

Procedure:

A. To measure the horizontal angle ABC, see Fig. 2.27.

1. Set up the Theodolite on station B and level it.
2. Keep the vertical circle at left (face left case).
3. Release the upper clamp and turn upper plate till the zero of any vernier (say A) of upper plate coincides with the zero of main scale of lower plate. Tighten the clamps of both the plates and co-incide the two zeros exactly by turning the upper tangent screw. Take the readings of both verniers. The readings of verniers A and B would be 0° and 180° respectively.
4. Release the lower clamp and sight the station A by swinging the telescope. Tighten the lower clamp and bisect the station point A accurately by lower tangent screw.
5. Release the upper clamp and swing the telescope in clockwise direction and sight the station point Q. Tighten the upper clamp and bisect the point C accurately by upper tangent screw.
6. Take readings of both verniers. The reading of vernier A gives the angle ABC directly while reading of vernier B gives the angle ABC by deducting 180°. Take average of the two values of angle ABC.
7. Keep the vertical circle at right by transiting the telescope (face right case) and repeat the above process and determine the angle ABC.
8. Take mean value of angle ABC obtained by steps (6) and (7).

Note: The zero of the vernier is initially set on zero of main scale for convenience purpose only, it can be set at any convenient reading.

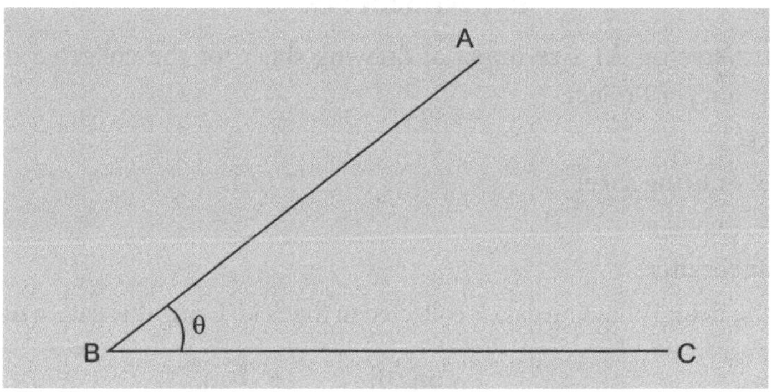

Fig. 2.27: Determination of horizontal angle

Results: The measured horizontal angle is ________.

B. To measure the vertical angle ABC, see Fig. 2.28.

1. Set the instruments at the station B and level it with the plate level
2. Release the vertical clamp and rotate the telescope in vertical plane till zero of any vertical vernier coincides with the zero of vertical circle. Tighten the vertical clamp and coincide the two zeros exactly by turning the vertical tangent screw. Check bubble of altitude level, if it is not in centre, bring it in centre with the help of the clip screw.

3. Release the clamp of vertical circle and rotate the telescope in vertical plane to sight the object A. Tighten the vertical clamp and use vertical tangent to bisect the object accurately.
4. Read both the vertical verniers C and D. The mean of readings of two verniers will give the value of the required angle
5. Change the face and repeat the process to get another value of required angle
6. Take average of the two values of required angle obtained in steps (4) and (5).

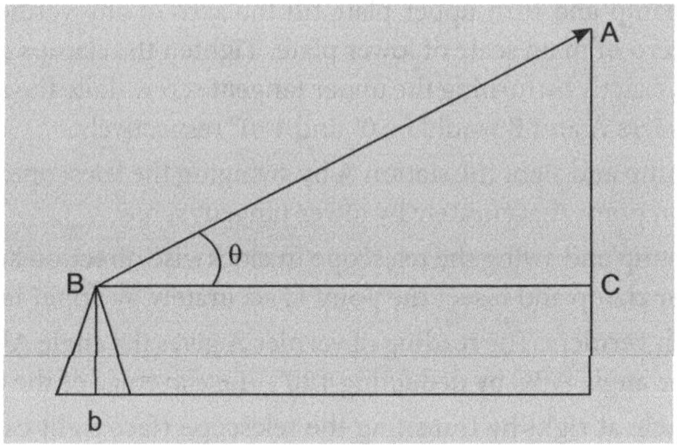

Fig. 2.28: Determination of vertical angle

Results: The measured Vertical angle is ________.

Experiment No. - 2

Objective: Plot the traverse on A1 size imperial drawing sheet for the collected data from preceding Theodolite Survey Project.

Required Accessories:

1. A1 size imperial drawing sheet
2. Notebook, scale, pencil etc.
3. Protector or mini drafter

Collected data from the field: Following data is collected in the field. Using this data, a traverse is to be plotted

Line	Length	Fore Bearing
AB	L_1	θ_1
BC	L_2	θ_2
CD	L_3	θ_3
DE	L_4	θ_4
EA	L_5	θ_5

Procedure: See Fig. 2.29.

1. Draw magnetic north at A, Draw a line at A setting F.B. as θ1 by protector
2. Mark point B on this line at length L1 choosing some scale
3. Draw magnetic north at B, Draw a line at B setting F.B. as θ 2 by protector
4. Mark point C on this line at length L2 choosing some scale
5. Similarly draw lines CD, DE, and EA
6. See weather line EA ends at A, if not draw closing error AA'
7. Adjust closing error by Bowditch's rule or transit rule.

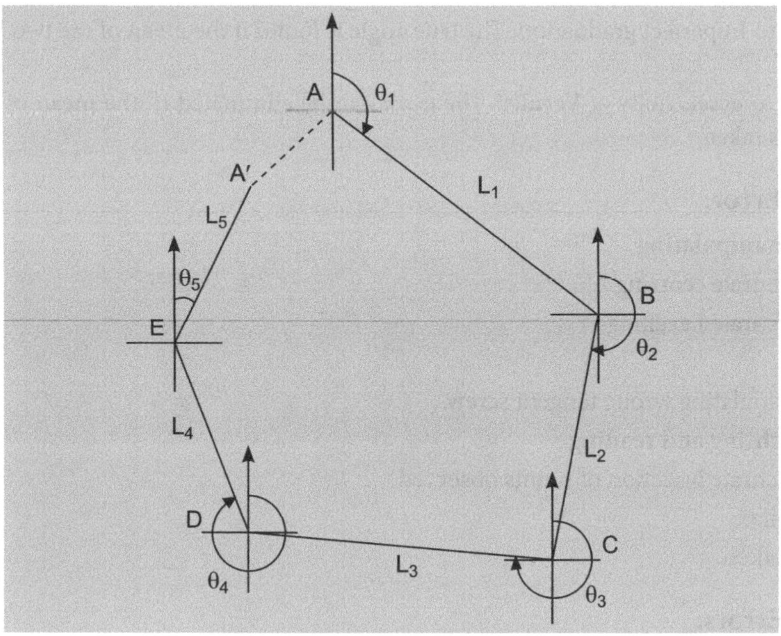

Fig. 2.29: Plotting of Traverse

KNOW MORE

Sources of errors in theodolite work

(i) Instrumental error
(ii) Personal error
(iii) Natural error

(i) Instrumental

(a) Error due to imperfect adjustment of plate levels: The error may be eliminated by levelling carefully with respect to the altitude bubble if it is in adjustment

(b) Error due to line of collimation is not perpendicular to the horizontal axis: The error can be eliminated if both face observations are taken

(c) Error due to horizontal axis is not perpendicular to the vertical axis: The error can be eliminated if both face observations are taken

(d) Error due to the axis of telescope level and line of collimation are not parallel: The error can be eliminated if both face observations are taken

(e) Error due to faulty adjustment of the vertical circle Vernier: The error can be eliminated if both face observations are taken

(f) Error due to eccentricity of inner and outer axes: The true angle is found if the mean of the two Vernier readings is taken.

(g) Error due to Imperfect graduation: The true angle is found if the mean of the two Vernier readings is taken.

(h) Error due to eccentricity of Vernier: The error can be eliminated if the mean of the two Vernier readings is taken.

(ii) Personal Error:

(a) Error in manipulating

(i) Inaccurate centring

(ii) Inaccurate Levelling.

(iii) Slip.

(iv) Manipulating wrong tangent screw.

(b) Errors in sighting and reading.

(i) Inaccurate bisection of points observed.

(ii) Parallax.

(iii) Mistakes.

(iii) Natural Errors:

(a) Unequal atmospheric refraction due to high temperature.

(b) Unequal expansion of parts of telescope and circles due to temperature changes.

(c) Unequal settlement of tripod .

(d) Wind producing vibrations.

History:

Prior to the present theodolite, instruments like the groma, geometric square, dioptra, and other graduated circles (circumferentor) and semicircles (graphometer) were used to find horizontal and vertical angles

First instrument was developed in 1725 by Jonathan Sisson. This instrument had a telescope, the base plate with spirit levels, compass and adjusting screws. The graduated circles were read by a vernier scale.

Some theodolites are shown with the time of development

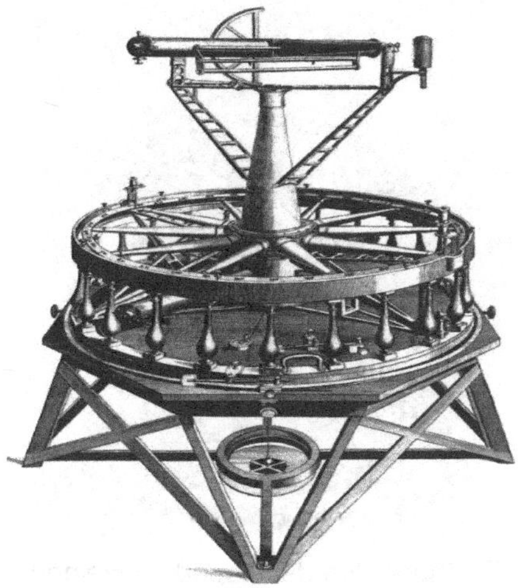

3

Fig. 2.30: Jesse Ramsden's Great Theodolite of 1787

Fig. 2.31: A theodolite of 1851

Fig. 2.32: A transit theodolite with six-inch circles, manufactured in Britain in 1910 by Troughton & Simms

Fig. 2.33: Wild T2 theodolite developed by Heinrich Wild in 1919

REFERENCES AND SUGGESTED READINGS

1. B.C Punmia, Surveying Vol –II, Laxmi Publication (P) Ltd. , 2023.
2. N.N. Basak, surveying & Levelling,McGraw hill Iindia, Private Limited, Noida, 2017.
3. R. Agor, A Textbook of Surveying & Levelling, Khanna Publication, 2015.
4. Bhavikatti SS, Surveying and Levelling Vol. II, I.K International Publishing House Pvt. Ltd, 2016.
5. S.K. Duggal, Surveying Vol. I, McGraw Hill Publishing Company Ltd., 2013.
6. C. Venkatramaiah, Textbook of Surveying, University Press (India) Limited, 2011.
7. Saikia MD, Das BM, Das MM, Surveying, Prentice Hall India Learning Limited, 2010.
8. T.P. Kanetkar and S.V. Kulkarini, Surveying and Levelling, Vidyarthi Griha Prakashan, 1967,
10. Link for NPTEL Vedio - https://youtu.be/6d4mERJFPpI

DYNAMIC QR CODE FOR FURTHER READING

Notes (Theodolite survery).png

Introduction to theodolite.png

Latitude and departure (1).png

Lecture by, IIT Kanpur.png

Theodolite traversing.png

3 Tacheometric Surveying

UNIT SPECIFICS

Through this unit we have discussed the following aspects:

- Principle of Tacheometry
- Tacheometer and its components including Anallatic lens
- Field method for determining tacheometer constants
- Methods of determining vertical and horizontal distances by tacheometer
- Limitations of Tacheometry

RATIONALE

This unit explains the principle of tacheometry. The tacheometer and its components are described. The utility of Anallatic lens is discussed. Tacheometric formula for determining horizontal distance with line of sight horizontal and staff vertical is derived. The field method of determining constants of tacheometer is presented.

The fixed hair method with staff held vertical is derived and explained to determine the horizontal and vertical distances by tacheometer. The limitations of tacheometry are discussed.

PRE-REQUISITE

Mathematics: Basic knowledge of Trigonometry

Basic Surveying (Third Semester Diploma)

UNIT OUTCOMES

List of outcomes of this unit is as follows:

U3-O1: State the principle of tacheometry

U3-O2: Describe the tacheometer and its components

U3-O3: Derive the tacheometer formula for distance measurement

U3-O4: Apply field methods of tacheometry

U3-O5: List the limitations of tacheometry

Unit-3 Outcomes	EXPECTED MAPPING WITH COURSE OUTCOMES (1- Weak Correlation; 2- Medium correlation; 3- Strong Correlation)				
	CO-1	CO-2	CO-3	CO-4	CO-5
U3-O1	1	1	3	1	1
U3-O2	1	1	3	1	1
U3-O3	1	1	3	1	1
U3-O4	1	1	3	1	1
U3-O5	–	1	2	1	1

3.1 INTRODUCTION

For linear measurement, chaining is the most accurate method, hence widely used. However, on rough and undulated ground, chaining is difficult and does not provide good accuracy, it is also a slower process. When the ground is rough and the work is to be carried out speedily with ordinary precision, tacheometry may be a good choice.

The Tacheometry is a method of determining vertical and horizontal distances by optical means. It is fast and convenient method but provides less accuracy on flat ground as compared to chaining, however, gives comparable precision on rough and steep grounds. It is generally used to prepare contours on a topographic map. It also provides a check on the distances measured by chaining.

3.2 PRINCIPLE OF TACHEOMETRY

The principle of tacheometry is derived from the property of an isosceles triangle that the ratio of the distance of the base from the vertex and the base length is always constant in an isosceles triangle.

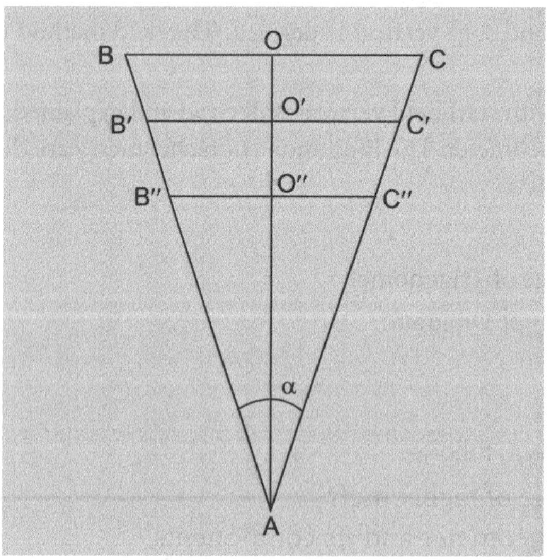

Fig. 3.1: Principle of Tacheometry

In Fig. 3.1, ΔABC, ΔA'B'C' and ΔA"B"C" are isosceles triangles hence,

$$\frac{AO}{OC} = \frac{AO'}{O'C'} = \frac{AO''}{O''C''} = \frac{\cot\alpha}{2} = K$$

where K is constant wholly depends on apex angle α.

This principle is used to relate staff intercept, stadia hair interval and ground distances in the tacheometry.

3.3 INSTRUMENTS USED IN TACHEOMETRY

3.3.1 Tacheometer

It is vernier transit theodolite fitted with stadia diaphragm. It consists of three horizontal hairs, one is in centre and other two are at equidistant from central hair at the top and the bottom. Fig. 3.2 shows different types of stadia diaphragm. The additive constant and multiplying constant should be zero and 100 respectively. This is achieved by providing anallatic lens in the telescope of the tacheometer. The anallatic lens is discussed later. The powerful telescope should be provided with 20 to 30 times magnification. The telescope objective lens should be provided with an aperture of about 40 mm diameter.

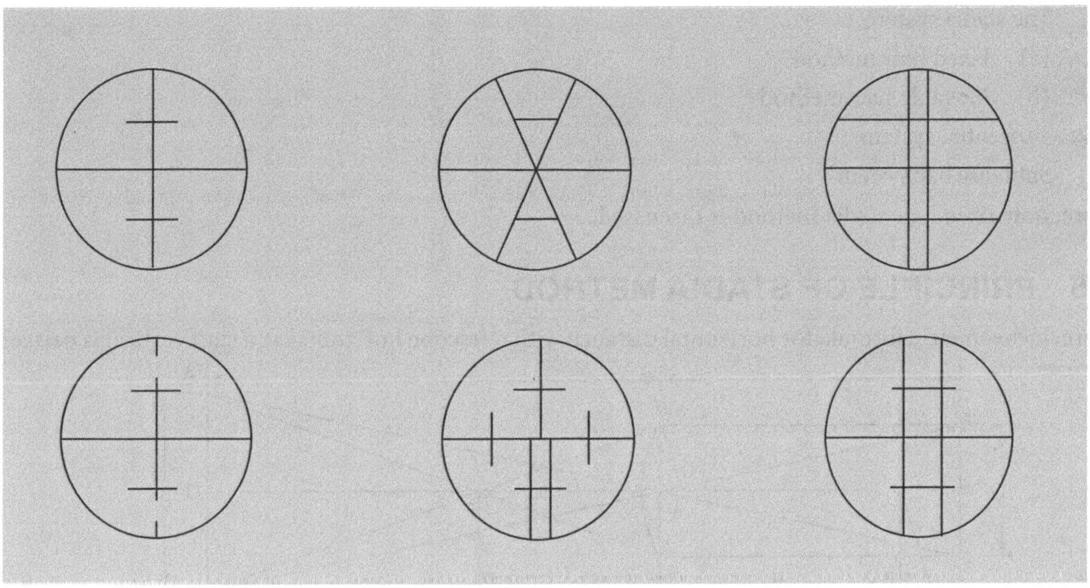

Fig. 3.2: Patterns of stadia diaphragm

3.3.2 Stadia Rod

For short distances a simple levelling staff with 5 mm least count may be used. For large distances a special large rod known as stadia rod, is used. It is one piece, 3 to 5 meter long rod. The graduation in the rod should be bold having width between 50 to 150 mm. (Fig. 3.3).

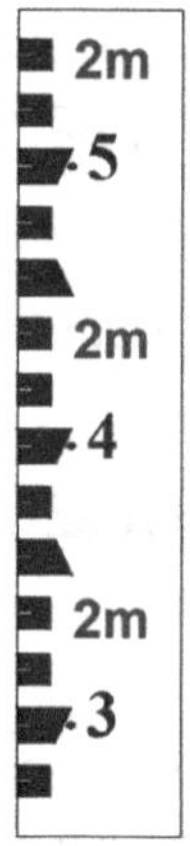

Fig. 3.3: Stadia Rod

3.4 SYSTEMS OF TACHEOMETRIC MEASUREMENT

The system of tacheometric measurement is classified as under:

1. The stadia system
   (a) Fixed hair method
   (b) Movable hair method
2. Tangential system
3. Subtense bar system

Here, only fixed hair stadia method is discussed.

3.5 PRINCIPLE OF STADIA METHOD

Here, tacheometric formula for horizontal distance with telescope horizontal and staff vertical is derived

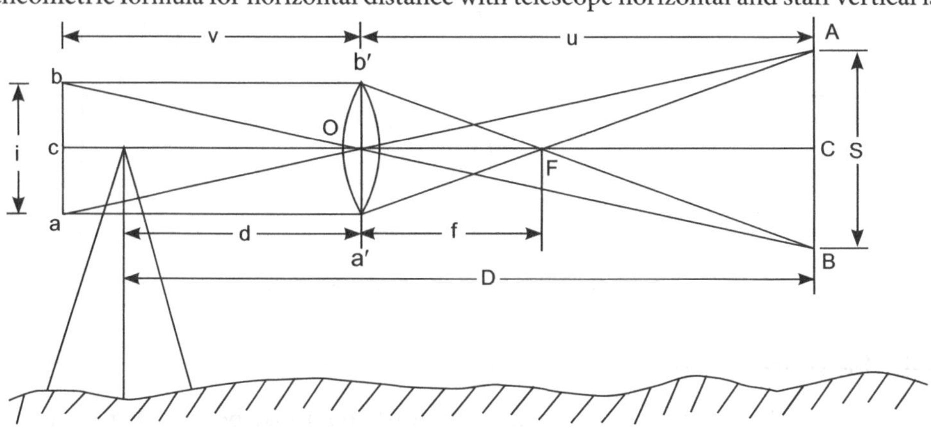

Fig. 3.4: Principle of Stadia Method

Refer Fig. 3.4

Here, O is the object lens of external focussing telescope,

a, b and c are three horizontal hairs of diaphragm,

Let i = stadia hair interval ab

s = staff intercept AB corresponding to stadia hair interval ab

f = focal length of the objective

u = horizontal distance between staff and optical center O of the objective

v = horizontal distance between cross hairs and optical center O of the objective

d = horizontal distance between vertical axis of the instrument and optical centre of the objective

D = horizontal distance between vertical axis of the instrument and staff

From similar Δs, aOb and AOB

$$\frac{OC}{Oc} = \frac{AB}{ab}$$

OR

$$\frac{u}{v} = \frac{s}{i} \quad ...(i)$$

u and v are conjugate focal length distances of objective

hence, by standard formula

$$\frac{1}{f} = \frac{1}{u} + \frac{1}{v} \quad ...(ii)$$

Multiply fu throughout

Putting from $\frac{u}{v} = \frac{s}{i}$ from Eqn (i)

but $D = u + d$

or $D = \frac{fs}{i} + (f + d)$

$$D = ks + C \quad (3.1)$$

This is called tacheometric distance equation in which

is known as multiplying constant

And C = f + d is known as additive constant for the instrument.

Alternate Method:

Refer Fig. 3.4

From similar Δs a'b'F and ABF

$$\frac{CF}{OF} = \frac{AB}{a'b'}$$

OR

$$CF = \frac{AB}{a'b'} \times OF$$

OR

$$CF = \frac{f}{i} . s$$

But $$D = CF + f + d$$

Putting

$$CF = \frac{f}{i} . s$$

$$D = \frac{fs}{i} + (f + d)$$

$$D = ks + C \qquad \text{same as eqn. ... (3.1)}$$

3.6 FIELD METHODS FOR DETERMINING CONSTANTS OF TACHEOMETER

First Method:

In this method, the additive constant C= (f+d) is measured directly along the height of the telescope while the multiplying constant K is determined through field measurement. Following steps are followed

1. Sight the telescope to a distant object and focus for clear image. Measure the distance along the telescope between the objective and the plane of cross hair.

   In relation, $$\frac{1}{f} = \frac{1}{u} + \frac{1}{v},$$

   u is very large, hence focal length f is approximately equal to v, i.e., f is equal to distance between the objective and the plane of cross hair.

2. Measure the distance between objective and vertical axis of instrument. This distance is equal to d. Thus, the additive constant C = (f+d) is known

3. Put the staff at a known distance D1 and note the corresponding staff intercept s_1. Using the equation 3.1 and putting the values of D, C and s corresponding this observation

   $$D_1 = ks_1 + C$$

   gives, $$k = \frac{D_1 - C}{s_1}$$

4. Similarly for different values of D say D_2, D_3, etc., corresponding staff intercepts s_2, s_3 are determined and more values of K is calculated. And a mean value of multiplying constant K is found out.

Second Method:

In this method, following steps are followed:

1. Measure a line PQ, say 100 m long, on levelled ground and insert pegs at 25 m intervals.

2. Set the instrument at P and obtain staff intercepts s_1, s_2, s_3 and s_4 on staff held vertical at each peg keeping horizontal line of sight
3. Form number of simultaneous equations by putting different values of D and corresponding values of s in distance equation D = ks + C.
4. Solve simultaneous equations in pair to get the values constants C and K and mean values are found out.

3.7 ANALLATIC LENS

It is a convex lens fitted between eyepiece and objective of telescope of the tacheometer. The use of this lens was invented by a scientist Porro in 1840. The purpose of this lens is to reduce the additive constant to zero.

In distance equation D = K.s + C or K.s = D – C, the staff intercept s is proportional to (D – C) i.e., distance between the staff and the exterior focus of the objective. By introducing anallatic lens , the vertex of the tacheometric triangle is brought at vertical axis of the instrument making staff intercept is directly proportional to horizontal distance D. The additive constant C = f + d, becomes zero, then

$$D = k.s \qquad \text{... (3.2)}$$

In above equation, by adjusting the setting in the instrument, the multiplying constant K is made as 100, so the equation 3.2 is reduced to

$$D = 100\ s$$

Hence multiplying staff intercept by 100, gives directly distance between staff station and instrument station.

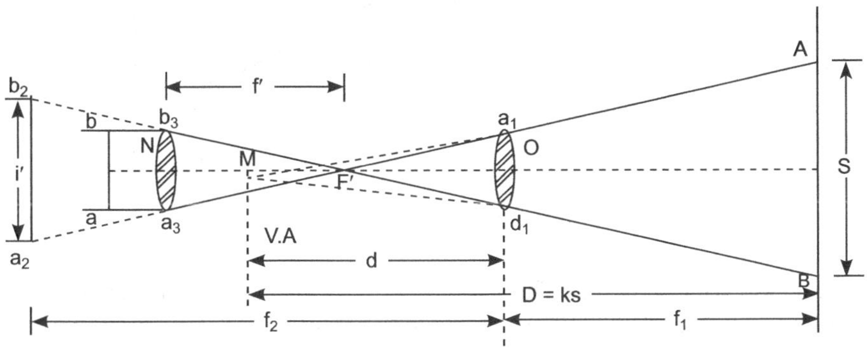

Fig. 3.5: Anallatic Lens

Advantage of Anallatic lens:

It simplifies the computations

Disadvantage of Anallatic lens:

(i) It increases the absorption of light in telescope resulting in reduction of brightness of image

(ii) It enhances the initial cost of the instrument.

(iii) This is possible only in external focussing telescopes

3.8 DETERMINATION OF HORIZONTAL AND VERTICAL DISTANCES BY TACHEOMETER

In this section, distance and elevation formulae with inclined line of sight (at angle of elevation and angle of depression) with staff held vertical, using fixed hair method, are derived.

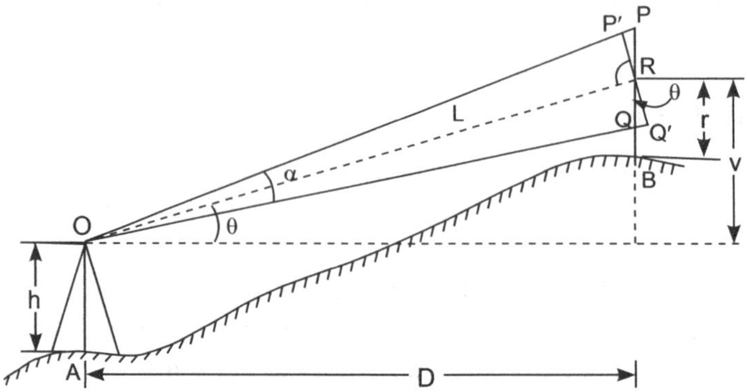

Fig. 3.6: Elevated line of sight (Vertical holding of staff)

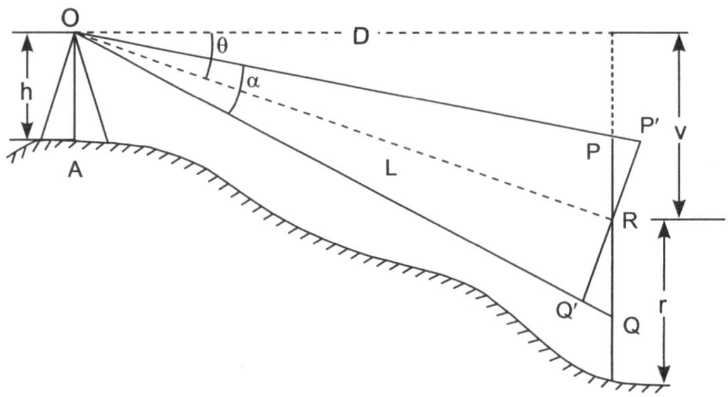

Fig. 3.7: Depressed line of sight (Vertical holding of staff)

Refer Fig. 3.6

The instrument is set at station A. The staff is held vertical at point B having horizontal distance D from P. The line of sight OR is inclined, makes an angle θ from horizontal. The staff intercept PQ is not perpendicular to line of sight OR but it is perpendicular to P'Q'.

Let R is midpoint of staff intercept corresponding to central hair reading.

V = Vertical distance between centre of instrument O and R

r = central hair reading

Thus, the angle between PQ and P'Q' is θ.

The ∠ PP'R = 900 +α/2 being the exterior angle of Δ P'OR

Similarly , from Δ ROQ', ∠ OQ'R = ∠ QQ'R = 900 – α/2

As α is very small, can be neglected, hence ∠ PP'R and ∠ QQ'R may be taken 90°

From Δ PRP', P'R = PR cosθ or P'Q' = PQ cosθ = s cosθ

The line P'Q' is particular to the line of sight OR, hence distance formula can be applied

$OR = L = k.\ P'Q' + C = K\ s.\ \cos\theta + C$

The horizontal distance $D = L \cos \theta$

Or

$$D = (k\ s \cos \theta + C) \cos \theta$$

Or

$$D = k\ s \cos2 \theta + C \cos \theta \tag{3.3}$$

Similarly,

$$V = L \sin \theta = (k\ s \cos \theta + C) \sin \theta$$

Or

$$V = k\ s \cos \theta\ .\ \sin \theta + C \sin \theta$$

Or

$$V = k\ s.\ \sin2\theta/2 + C \sin \theta(3.4)$$

The equations (3.3) and (3.4) are distance and elevation formulae for inclined line of sight with staff held vertical, using fixed hair method.

The Elevation of staff station B may be determined as under:

(i) For angle of elevation (Fig. 3.6)

Elevation of B = Elevation of A + h + V – r

(ii) For angle of depression (Fig. 3.7)

Elevation of B = Elevation of A + h - V - r

3.9 LIMITATION OF TACHEOMETRY

(i) It is less accurate than chaining.

(ii) It is an approximate method of distance and elevation measurement.

(ii) It is not used for work of high precision.

(iii) It is generally used for topographic work and as a check over the works done by chaining.

UNIT SUMMARY

- The principle of tacheometry is derived from the property of an isosceles triangle that the ratio of the distance of the base from the vertex and the base length is always constant in an isosceles triangle.
- Instruments used in Tacheometry
  - (i) Tacheometer
  - (ii) Stadia rod
  - (iii) Anallatic lens – it is used to reduce the additive constant to zero.

- Systems of tacheometric measurement
  - (i) The stadia system
    - (a) Fixed hair method
    - (b) Movable hair method
  - (ii) Tangential system
  - (iii) Subtense bar system
- Tacheometric distance formula

  $D = k.s + C$
- Distance and elevation formula of inclined line of sight.

  $D = k.s \cos^2 \theta + C \cos \theta$

  $V = k.s \sin^2 \theta/2 + C \sin \theta$

EXERCISES

Multiple Choice Questions

1. Which of the following is the example of an indirect method of surveying?
   (a) Tacheometry (b) Chain surveying
   (c) Contouring (d) All of the mentioned
2. The chain surveys being eliminated?
   (a) In Tacheometry (b) In Ranging
   (c) In Contouring (d) In Random line method
3. Tacheometer has ________ number of horizontal hairs.
   (a) 2 (b) 4
   (c) 3 (d) 5
4. Which of the horizontal hairs are used to determine staff intercept in stadia diaphragm of tacheometer?
   (a) Upper and central (b) Lower, central and upper
   (c) Upper and lower (d) Central and lower
5. In the D = Ks + C, where K is ________.
   (a) Multiplying constant (b) Additive constant
   (c) Divisional Constant (d) Subtractional Constant
6. For an anallactic lens, in $D = ks + C$, which of the following is zero?
   (a) D (b) k
   (c) s (d) C
7. The value of additive constant in anallactic lens?
   (a) 100 (b) 0
   (c) 0.3 (d) 0.4

8. The value of the additive constant lies in the range of ________.

   (a) 0.2 to 0.3 m (b) 0.5 to 0.6 m
   (c) 0.3 to 0.45 m (d) 0.6 to 0.7 m

9. If the ground is undulating, rough, inaccessible. In these circumstances ________ methods is used to obtain distances.

   (a) Direct methods (b) Chain surveying
   (c) Theodolite Method (d) Tacheometry

10. By using tacheometric methods, we can determine the elevations.

    (a) True (b) False

11. The purpose of the analectic lens in tacheometer.

    (a) Increase Magnification
    (b) Reduce effective length of telescope
    (c) To eliminate multiplying constant
    (d) To make the staff intercept proportional to its distance from tacheometer.

12. A stadia method in tacheometry is used to determine

    (a) Horizonal angles (b) Vertical angles
    (c) Horizontal Distance (d) Horizontal and vertical distances

13. A transist theodolite fitted with a stadia diaphragm and anallatic lens is known as

    (a) Transist theodolite (b) Tacheometer
    (c) Substense theodolite (d) Astronomical theodolite

14. The multiplying constant is denoted by

    (a) f / i (b) i / f
    (c) i × f (d) f + d

15. The additive constant is denoted by

    (a) f / i (b) i / f
    (c) i × f (d) f + d

ANSWERS															
1.	(a)	2.	(a)	3.	(c)	4.	(c)	5.	(a)	6.	(d)	7.	(b)	8.	(c)
9.	(d)	10.	(a)	11.	(d)	12.	(d)	13.	(b)	14.	(a)	15.	(d)		

Short and Long Answer Type Questions

Short Questions:

1. Explain principle of tacheometry.
2. Name components of Tacheometer.
3. What is Anallatic lens?
4. Name the various systems of tacheometric measurement.
5. Write tacheometric formula for horizontal distance with notations .

6. Write one method of determining the constants of tacheometer.
7. Write distance and elevation tacheometric formulae for fixed hair method having staff vertical with notations.
8. Write limitations of tacheometry.

Long Questions:

1. Describe various components of tacheometry
2. Derive tacheometric formula for horizontal distance using fixed hair method with horizontal line of sight.
3. Describe two methods of determining the constants of tacheometer
4. Derive distance and elevation tacheometric formulae for fixed hair method with staff held vertical having inclined line of sight.

Solved Numerical Problem

Numerical 1: The following reading were taken with a tacheometer on a vertical staff, Calculate tacheometric constant.

Horizontal Distance (m)	Stadia Reading (m)		
45	1.885	2.110	2.335
60	2.860	3.160	3.460

Solution: According to the distance formula

$$D_1 = Ks_1 + C \quad ...(1)$$

$$D_2 = Ks_2 + C \quad ...(2)$$

Putting the values in equation (1)

$$45 = K(2.335 - 1.885) + C$$

$$45 = K(0.45) + C \quad ...(3)$$

Putting the values in equation (2)

$$60 = K(3.460 - 2.860) + C$$

$$60 = K(0.6) + C$$

$$C = 60 - K(0.6)$$

Now put the value of C in equation (3)

$$45 = K(0.45) + 60 - K(0.60)$$

$$45 - 60 = -0.15\,K$$

$$-15 = -0.15\,K$$

$$K = 100$$

Put the value of K in equation (3)

$$45 = K(0.45) + C$$

$$45 = 100(0.45) + C$$

$$C = 0$$

Numerical 2: The stadia reading with horizontal sight at a vertical staff held 55 m away from the tacheometer were 1.950 and 2.110. The focal length of the object glass was 26 cm. The distance between the object glass and trunion axis of a tachometer was 16 cm. Calculate the stadia interval.

Solution: We know that

Horizontal distance (D) = KS + C

$$D = (f/i) * S + (f + d) \quad ...(1)$$

Here,

$$D = 50 \text{ m}$$

$$S = 2.110 - 1.950 = 0.16$$

$$f = 26 \text{ cm} = 0.26 \text{ m}$$

$$d = 16 \text{ cm} = 0.16 \text{ m}$$

Substitute above values in equation no (1)

$$55 = ((0.26 \times 0.16) / i) + (0.26 + 0.16))$$

$$i = 0.00083 \text{ m}$$

$$i = 0.83 \text{ mm}$$

Numerical 3: A staff held vertically at 50 m and 100 m from the centre of the theodolite with a stadia hair, the staff intercept is 0.500 and 1.000 respectively. The instrument was setup at a station P of RL 1950.95 m and the total height of instrument was 1.5 m. The hair reading on a staff held vertically at station Q were 2.050, 2.900 and 3.750 with the line of sight is horizontal. Calculate the horizontal distance of PQ and RL of Q pint.

Solution: (i) Determination of Constants (K & C)

According to the distance formula,

$$D_1 = Ks_1 + C \quad ...(1)$$

$$D_2 = Ks_2 + C \quad ...(2)$$

Putting the values in equation (1) & (2)

$$50 = K\,(0.500) + C \quad ...(3)$$

$$100 = K\,(1.000) + C \quad ...(4)$$

From equation (3)

$$C = 50 - 0.500\,K \quad ...(5)$$

Put the value of C in Equation (4)

$$100 = 1.000\,K + 50 - 0.500\,K$$

$$K = 100$$

Now put the value of K in equation (5)

$$C = 50 - 0.500\,(100)$$

$$C = 50 - 50$$

$$C = 0$$

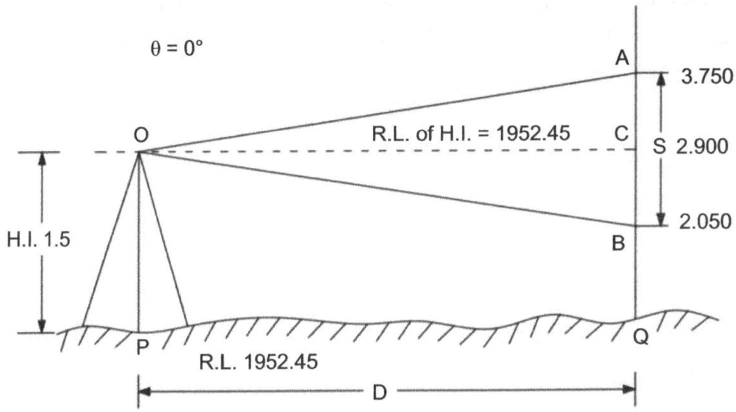

Fig. 3.8: Determination of horizontal distance and R.L.

(ii) Determination of distance & R.L.

s = 3.750 – 2.050 = 1.700 m

K = 100 & C = 0 (Determined above)

Put the values in equation of horizontal distance

D = KS + C

D = 100 (1.700) + 0

D = 170m

Calculation of RL of Q point

RL of Q = R.L. of P + H.I. – Reading of central hair

RL of Q = 1950.95 + 1.5 – 2.900

= 1949.55 m

Numerical 4: A tacheometer has a diaphragm with three cross hairs spaced at distance 1.20 mm. The focal length of the object glass is 24cm and the distance from the object glass to the trunnion axis is 11cm. Calculate the tacheometric constant.

Solution: f = 24 cm

d = 11 cm

i = 2 × 1.20 = 2.40 mm = 0.24 cm

Substituting the value in formula

K = f/i = 24/0.24 = 100

and C = (f+d) = 24+11 = 35 cm

Multiplying constant (k) = 100

Additive constant (c) = 0.35

Numerical 5: A levelling staff is held vertical at the distance of 120 m and 290 m from the axis of tacheometer and the staff intercept for horizontal sights are 1.64 m and 3.35 m metre respectively. Find the instrument constants.

The tacheometer is set up at station A and the staff is held vertical at a point B and the line of sight of the telescope inclined at an angle of depression of 10° to the horizontal, the readings on the staff are 3.670, 2.835, 2.000 m. Calculate its horizontal distance from A and the RL of B. The height of instrument is 1.5 m and RL of A is 450.5 m.

Solution: Given data

$$D_1 = 120 \text{ m}$$
$$s_1 = 1.64$$
$$D_2 = 290 \text{ m}$$
$$s_2 = 3.35$$

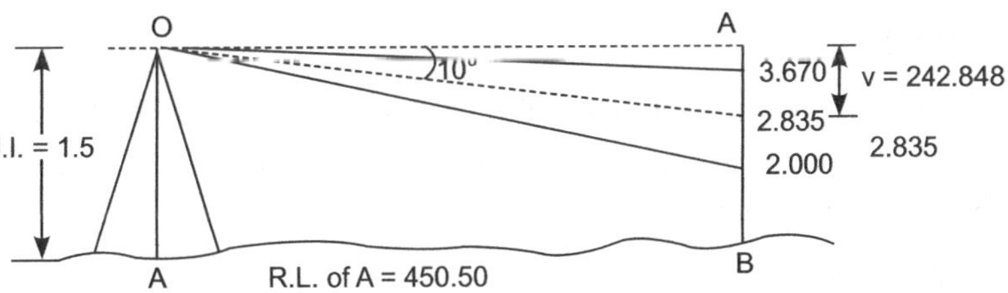

Fig. 3.9: Determination of horizontal distance and R.L.– Inclined sight

According to the distance formula,

$$D_1 = Ks_1 + C \quad ...(1)$$
$$D_2 = Ks_2 + C \quad ...(2)$$

Putting the values in equation (1)

$$120 = (K \times 1.64) + C \quad ...(3)$$

Putting the values in equation (2)

$$290 = (K \times 3.35) + C \quad ...(4)$$

From equation (3)

$$C = 120 - 1.64 \text{ K}$$

Putting the value of C in equation (4)

$$290 = 3.35 \text{ K} + (120 - 1.64 \text{ K})$$
$$290 - 120 = 3.35 \text{ K} - 1.64 \text{ K}$$
$$170 = 1.71 \text{ K}$$
$$K = 99.41$$

Putting the value of K in equation (4)

$$290 = 99.41 \times 3.35 + C$$
$$C = 0.8708$$

Hence, the constants of instruments are 99.41 & 0.8708

Now, $S = 3.670 - 2.000 = 1.670$ and $\theta = 10° \, 00'$

Horizontal distance = $Ks \cos^2 \theta + C \cos \theta$

$= 99.41 \times 1.670 \times \cos^2 10° 00' + 0.8708 \times \cos 10° 00'$

$= 82.67$ m

Vertical distance:

$V = Ks/2 \sin 2\theta + C \sin \theta$

$= 99.41 \times 1.670/2 \sin 20° + 0.8708 \sin 10°00'$

$= 242.848$m

R.L. of B = R.L. of A + H.I. – v – r

$= 450.5 + 1.5 - 242.848 - 2.835 = 206.317$ m

Unsolved numericals:

Numerical 1: A tachometer has a diaphragm with three cross hairs spaced at distance 1.25 mm. The focal length of the objective is 25 cm and the distance from the object glass to the trunnion axis is 11 cm. Calculate the tacheometric constant.

Solution: Multiplying constant = 100

Additive constant = 0.36

Numerical 2: A tacheometer is set up at an intermediate point on a traverse course AB and the following observations were staff held vertically.

Staff	Vertical angle	Staff intercept	Axial hair readings
A	+9°30'	2.250	2.105
B	+6° 00'	2.055	1.875

The instrument is fitted with an anallactic lens. The multiplying constant is 100. Compute the length AB and reduce level of B. if the reduced level of A is 350.56 m.

Solution: AB = 422.13 m,

R.L. = 352.56 m

Numerical 3: A tacheometer was set up at a station A and the readings on a vertically held staff were 2.255 and 2.605, the line of sight being at an inclination of +8° 24'. Other observation on the vertically held staff at bench mark provided the reading 1.640, 1.920 and 2.200, The inclination of the line of sight being +1°6'. Calculate the horizontal distance between A & B, and the elevation of the B is 418.685 m. The constant of the instrument is 100 and 0.3.

Solution: Elevation of collimation = 419.529 m

D = 68.80 m

R.L. at B = 427.079 m

Numerical 4: The stadia readings with horizontal line of sight on a staff held vertical at a point 55 m away from a tacheometer were 1.750 and 1.280. The focal length of the objective was 30 cm. The distance between the objective and trunnion axis of the tacheometer was 17 cm.

Solution: i = 2.85 mm

PRACTICAL

Objective: Use Thedolite as a Tacheometer to compute reduced levels and horizontal distances.

Instruments and Accessories Required: Tacheometer, Tripod Stand, Levelling Staff, Pegs, Arrows, etc.
Tacheometric constants: k = 100, c = 0

Theory: Several commonly used optical surveying instruments can be used for the purpose of tacheometry. A theodolite is the most preferred choice though. The only requirement for a theodolite to be a tacheometer is that the diaphragm of its telescope should possess upper and lower stadia hairs.

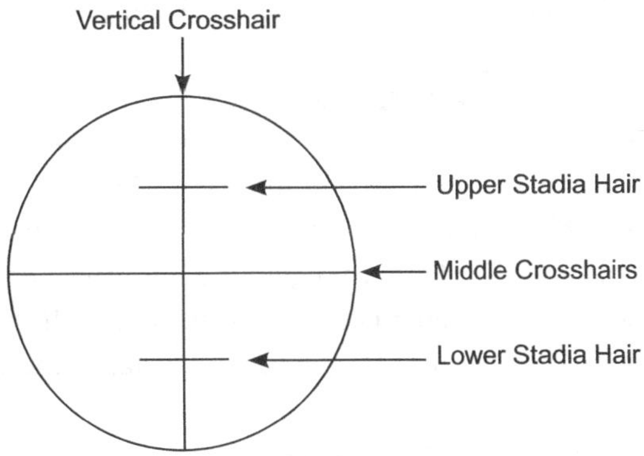

Fig. 3.10: Stadia Cross Hairs

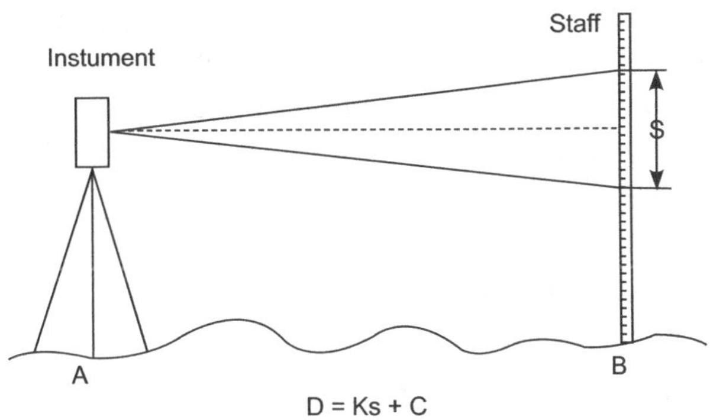

Fig. 3.11: Computation of horizontal distance and R.L.

Procedure:

Let the distance between two stations A and B is to be determined , also R. L. of B is to be determined, given R.L of A.

Steps to be followed to determine distance AB:

1. Establish the stations A and B
2. Set the theodolite at Station A
3. Make the line of sight of telescope horizontal setting the reading of vernier zero.
4. Place a levelling staff in the vertical position on station B.

5. Through the telescope sight this staff and properly focus the image using object focusing screw. Also make the stadia clearly visible by adjusting the eyepiece.
6. Note down the readings on levelling staff corresponding to upper, central and lower stadia hairs.
7. Find the staff intercept, s by finding the difference between upper and lower stadia hairs readings.
8. Calculate the distance by using the tachometric distance equation D = Ks + C Since the constants are known to us (K=100, C= 0), D = 100s.
9. Determine the H.I. by measuring the vertical distance from ground to the centre of the instrument at station A .
10. Determine the R. L. of the point B by following equation

    R. L. of station B = R. L. of station A + H. I. at A – central hair reading

KNOW MORE

- **Movable hair method** - In this method the staff intercept s remains fixed while the tacheometric angle α changes with the staff position. This can be achieved by sighting a graduated staff having the targets at some fixed distance apart and changing the interval i between the stadia hairs till the lines of sights corresponding to the stadia hairs bisect the targets. If the staff position is changed , the value of i also changed. The same expression for the distance measurement is used as for the stadia method.
- **Tangential method** - In this method, the horizontal and vertical distances between the tacheometer station and the staff station are found by observing verticals angles to the vanes fixed at a constant distance apart on the staff. Hence the stadia hairs are not used here, and the vane is sighted every time at the axial hair. Thus, two vertical angles are measured at each vane.
- **Direct method of tacheometer** - There are some direct reading instruments in tacheometry which reduce the computation time for measuring vertical and horizontal distances. The some instruments are mentioned below:
    1. Beaman Stadia arc
    2. Jeffcott Direct Reading Tacheometer
    3. Szepessy Direct Reading Tacheometer
- **Errors in tacheometry:** Error in tacheometric survey are as follows:
    1. **Instrumental Errors:** Instrumental error in tacheometric survey are as follows:
        (i) Imperfect adjustment of the tacheometer.
        (ii) Incorrect divisions on the stadia rod.
        (iii) Faulty value of the multiplying constant (f/t).
    2. **Personal Errors:** Personal error in tacheometric survey are as follows:
        (i) Wrong centering and faulty levelling of the instrument.
        (ii) Non-verticality of the staff or rod. It may be rectified by using a plumb bob.
        (iii) Wrong Focusing.
        (iv) Reading at wrong hair.

3. **Natural Errors:** Natural error in tacheometric survey are as follows:
   (i) High wind
   (ii) Unequal refractions
   (iii) Unequal expansion
   (iv) Bad visibility

REFERENCES AND SUGGESTED READINGS

1. B.C Punmia, Surveying Vol –II, Laxmi Publication (P) Ltd. , 2023.
2. N.N. Basak, surveying & Levelling,McGraw hill Iindia, Private Limited, Noida, 2017.
3. Bhavikatti SS, Surveying and Levelling Vol. II, I.K International Publishing House Pvt. Ltd, 2016.
4. R. Agor, A Text Book of Surveying & Levelling, Khanna Publication, 2015.
5. S.K. Duggal, Surveying Vol. I, McGraw Hill Publishing Company Ltd., 2013.
6. C. Venkatramaiah, Textbook of Surveying, University Press (India) Limited, 2011.
7. Saikia MD, Das BM, Das MM, Surveying, Prentice Hall India Learning Limited, 2010.
8. T.P. Kanetkar and S.V. Kulkarini, Surveying and Levelling, Vidyarthi Griha Prakashan, 1967,
9. Link for NPTEL Video – https://youtu.be/cjTT4a-dg5s

Dynamic QR Code for Further Reading

Introduction to techeometer

Techeometric survey-(Video lecture)

Video lecture (NPTEL)

Method of techeometer

4 Curves

UNIT SPECIFICS

Through this unit we have discussed the following aspects:

- Purpose and types of Curves
- Elements of Curves
- The ways of designation of curves
- Setting of simple circular curves – Offsets from long chord method
- Setting of simple circular curves – Rankine's method of deflection angles

RATIONALE

This unit describes purpose and types of Curves used in highway and railway. The elements and terminology used in curves is discussed in the chapter. The ways of designating the curves are explained. The setting of simple circular curves by method of offsets from long chord and Rankine's method of deflection angles is described.

PRE-REQUISITES

Mathematics: Basic knowledge of Trigonometry

Basic Surveying (Third Semester Diploma)

UNIT OUTCOMES

List of outcomes of this unit is as follows:

U4-O1: State the purpose and types of Curves.

U4-O2: List the elements of Curves.

U4-O3: Explain the ways of designation of Curves.

U4-O4: Apply the method of offsets from long chord for setting simple circular curve.

U4-O5: Apply the Rankine's method of deflection angles for setting simple circular curve.

Unit- 4 Outcomes	EXPECTED MAPPING WITH COURSE OUTCOMES (1- Weak Correlation; 2- Medium correlation; 3- Strong Correlation)				
	CO-1	CO-2	CO-3	CO-4	CO-5
U4-O1	-	1	1	1	1
U4-O2	-	1	1	1	1
U4-O3	-	1	1	1	1
U4-O4	-	1	1	1	1
U4-O5	-	1	1	1	1

4.1 INTRODUCTION

At intersection of roadways or railways, there is sharp change in direction. If a vehicle turns sharply at intersection, abrupt jerk at vehicle may takes place, even vehicle topple down due to centrifugal force. To avoid this situation gradual change in direction is accomplished at the intersection by introducing the curves. The intersection of roadways or railways may be in horizontal or vertical plane accordingly horizontal or vertical curves are introduced. In this chapter horizontal simple circular curve is presented.

4.2 CLASSIFICATION OF CURVE

Curves may basically be classified in two ways:

(i) Horizontal Curve
(ii) Vertical Curve

As stated above horizontal and vertical curve lie in horizontal and vertical plane respectively. The horizontal curve are further classified as under:

(i) simple circular curve
(ii) compound curve
(iii) reverse curve
(iv) transition curve

A **simple curve** is a single arc of the circle and it is tangential to both straights forming intersection (Fig. 4.1).

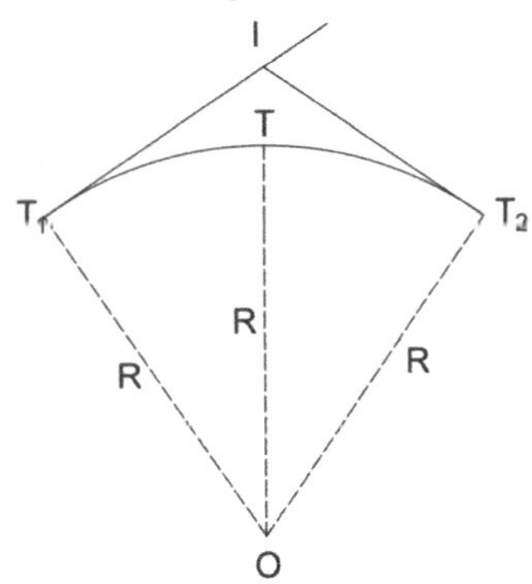

Fig. 4.1: Simple Circular Curve

A **compound curve** comprises of two or more simple circular curves of different radii having turn in same direction and meeting at common tangent point (Fig. 4.2).

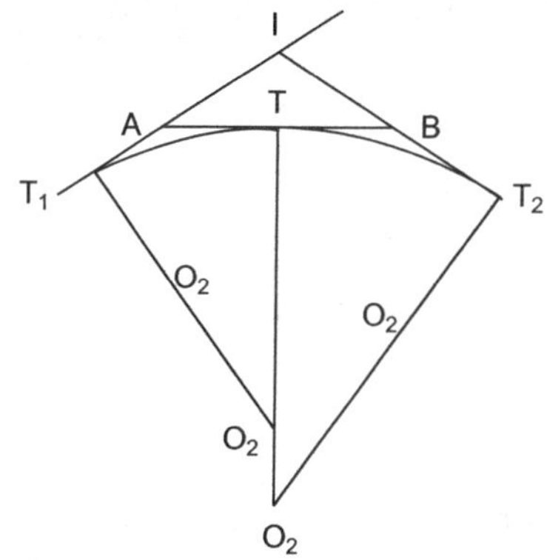

Fig. 4.2: Compound Curve

A **reverse curve** comprises of two simple circular curves of equal or different radii with their centres to the different sides of the common tangent point. It is rarely used in highway or railway but commonly used in railway yards (Fig. 4.3).

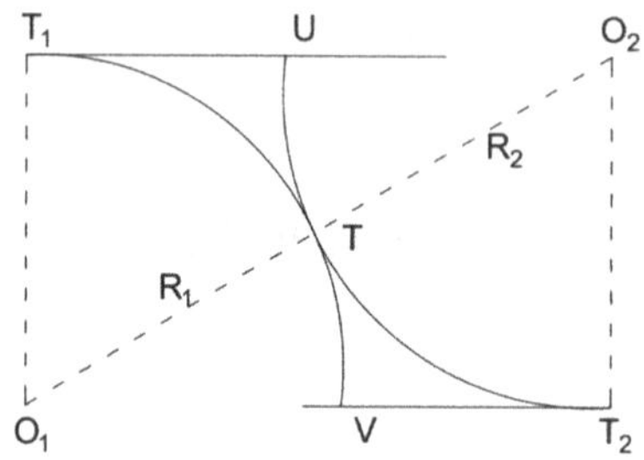

Fig. 4.3: Reverse Curve

Transition curve is used between a simple circular curve and a straight or introduced between two simple circular curves. The radius of transition curve changes gradually from finite to infinite or vice versa. It is widely used on roads or railway tracks as it provides a comfort ride (Fig. 4.4).

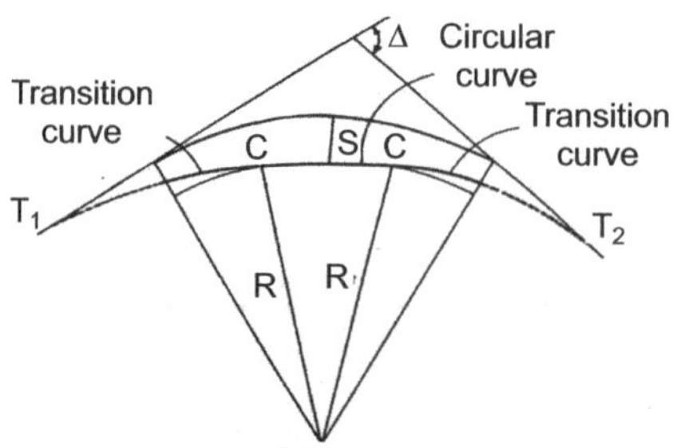

Fig. 4.4: Transition Curve

4.3 SIMPLE CIRCULAR CURVE

4.3.1 Elements of Circular Curves

Fig. 4.5 presents a simple circular curve. The basic elements of simple circular curve are as under:

(i) Back tangent (PT_1): It is tangent at the initial point of the curve , T_1.

(ii) Forward tangent (T_2Q): It is tangent at the last point of the curve, T_2.

(iii) Point of curve, P.C. (T_1): It is the starting of the curve where a straight converts to the curve.

(iv) Point of tangency, P.T. (T_2): It is the end of the curve where the curve converts to a straight.

(v) Point of intersection, P.I. (V): It is the intersection of back tangent PT_1 and forward tangent QT_2.

(vi) Deflection angle (Δ): It is the angle by which straight PV deflects to straight VQ.

(vii) Tangent distance or length (T_1V and VT_2): It is the distance between point of intersection and point of tangency or point of curve.

It is given by, T_1V and $VT_2 = OT_1 \tan \Delta/2 = R \tan \Delta/2$ ...(4.1)

(viii) External distance (VR): It is the distance between mid-point of the curve and point of intersection.

(ix) Long chord (T_1T_2): It is the distance between point of tangency and point of curve.

(x) Length of the curve ($T_1 RT_2$): It is the total length of the curve from point of tangency to point of curve.

(xi) Mid ordinate (RS): It is the distance between mid-point of the curve and mid-point of the long chord.

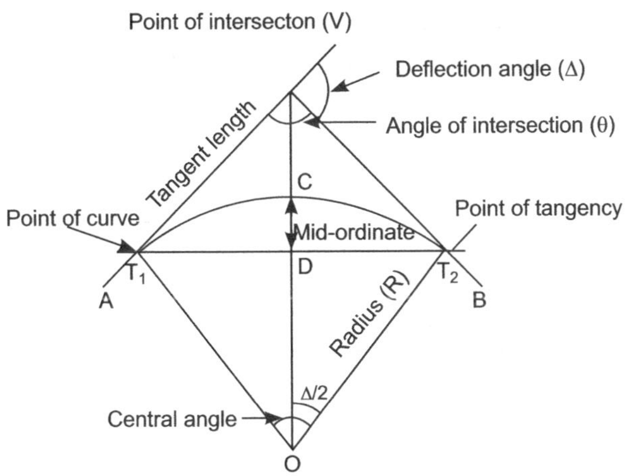

Fig. 4.5: Elements of Circular Curves

4.3.2 Designation of Curve

A simple circular curve is designated by following ways:

(i) By radius of arc

(ii) By degree of a curve

The degree of curve (D) may be defined as the angle subtended at the centre by an arc or angle subtended at the centre by a chord.

The arc or the chord length is taken a standard length, it is generally taken as one chain length.

The chain length may be 20 m or 30 m. Here it is taken as 30 m.

Relationship between radius and degree of curve:

By arc definition [Fig. 4.6 (a)]

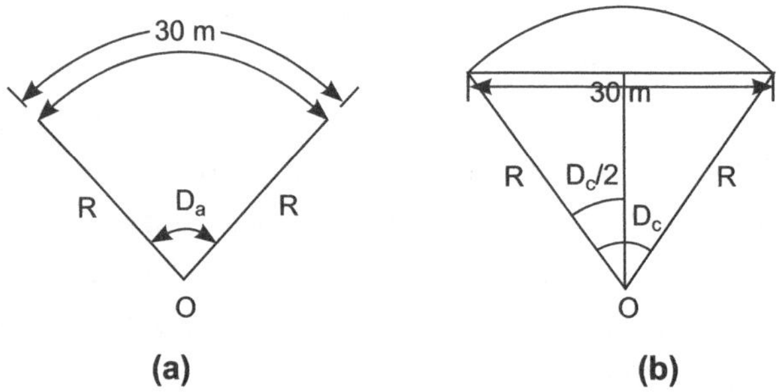

Fig. 4.6: Designation of curve

$$D_a = \frac{360°}{2\pi R} \times 30 = \frac{1718.88}{R} \quad ...(4.2)$$

By chord definition [(Fig. 4.6 (b)]

$$\sin\frac{D_c}{2} = \frac{15}{R}$$

$$\sin\frac{D_c}{2} = \frac{15}{R} \text{ radian, if } \frac{D_c}{2} \text{ is small}$$

Then
$$\frac{D_c}{2} = \frac{15}{R}$$

Converting in degree
$$\frac{D_c}{2} \times \frac{\pi}{180°} = \frac{15}{R}$$

$$D_c = \frac{360°}{\pi} \times \frac{15}{R}$$

$$= \frac{1718.88}{R}$$

Hence, for small value of degree of curve, both arc definition and chord definition give same result.

4.3.3 Methods of Setting Simple Circular Curves

The curves may be set out by following ways:

(i) **Linear methods:** This method is generally used when the lengths of curves are short and a high degree of precision is not needed.

(ii) **Angular methods:** In this method angle measuring instruments like theodolite is used with or without a chain/tape.

Before setting a curve, it is required to locate tangents, point of curve (P.C.), point of intersection (P.I.) and point of tangency (P.T.). First of all, point of intersection is established on the field and theodolite is set on it. The telescope of the theodolite is directed towards first straight and then it is transited through 1800 and Vernier readings are noted. Now telescope is directed towards second straight by turning it and vernier readings are noted. The deflection angle (Δ) is determined by difference of two verniers readings. The tangent length is calculated by following formula:

$$T = R \tan$$

The point of curve (T_1) and point of tangency (T_2) are established at tangent length distances from point of intersection (P.I.) by setting line of sight along the two straights.

Linear methods

Following are the various linear methods of setting out a simple circular curve:

(i) By offsets from the long chord

(ii) By perpendicular offsets from the tangent

(iii) By radial offsets from the tangent

(iv) By offsets from the chord produced (deflection method)

(v) By successive bisection of arcs

Here, by offsets from the long chord method is discussed

Offsets from the long chord method

Refer Fig. 4.7

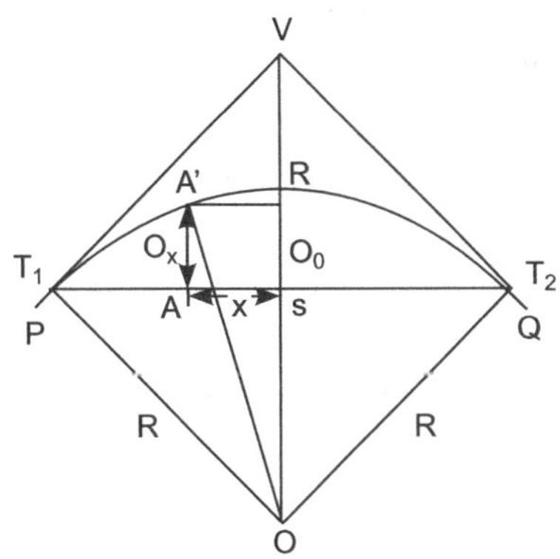

Fig. 4.7: Offsets from the long chord method

Let the curve T_1RT_2 to be set between two intersecting straights T_1V and T_2V. T_1 and T_2 are the point of curve and point of tangency respectively. R is the radius of the curve. The mid-ordinate is O_o and O_x is offset at point A at a distance x from the mid-point (S) of the long chord. L is the length of the long chord.

In Δ OST1,

$$OS = \sqrt{(OT_1^2 - ST_1^2)}$$

$$= \sqrt{(R^2 - (L/2)^2)}$$

$$RS = OR - OS$$

Or

$$Oo = R - \sqrt{(R^2 - (L/2)^2)}$$

In Δ OA'T,

$$OT = \sqrt{(R^2 - x^2)}$$

And,

$$OS = R - O_o$$

Then ordinate O_x at distance x from S is

$$O_x = AA' = OT - OS$$

$$O_x = \sqrt{(R^2 - x^2)} - (R - O_o) \qquad ...(4.3)$$

This is exact expression of O_x

$\sqrt{(R^2 - x^2)}$ can be approximated as under

$$\sqrt{(R^2 - x^2)} = R\,(1 - x^2/R^2)1/2$$

$$= R\,(1 - x^2/2R^2) + ...)$$

$$= R - x^2/2R$$

Neglecting higher terms,

Putting this value in equation (4.3)

$$O_x = R - x^2/2R - R + O_o$$

$$O_x = O_o - x^2/2R \qquad ...(4.4)$$

This is approximate expression of O_x.

The long curve is divided into even number of equal parts. The offsets at different intervals are calculated by putting various values of x in equation (4.3) or (4.4). The points at different offsets are established and joined to set out desired curve T_1RT_2.

Angular methods

Following are the various angular methods of setting out a simple circular curve:

1. Rankine's method of deflection angle
2. Two theodolite method
3. Tacheometric method

Here, Rankine's method of deflection angle, is discussed:

Rankine's method of deflection angle

This method is suitable for long curve with large radius. It gives fair results and used on highways and railways.

The deflection angle at any point on the curve is the angle at point of curve (P. C.) between the chord from P.C. to that point and the back tangent.

This method is based on following fact:

The deflection angle at any point is equal to half the angle subtended by the arc at the centre.

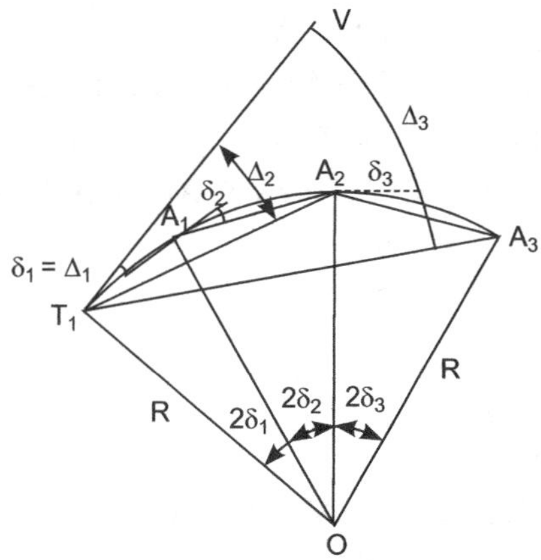

Fig. 4.8: Rankine's method of deflection angle

Refer Fig. No.- 4.8. Let T_1 is point of curve and A_1, A_2, A_3 etc. are the points on the curve, δ_1, δ_2, δ_3 etc. are the respective deflection angles between the chords and the respective tangents at T_1, A_1, A_2, etc. Δ_1, Δ_2, Δ_3 etc are the total tangential angle or deflection angles to the points A_1, A_2, A_3 etc.

C_1, C_2, C_3 etc. are lengths of the chords $T_1 A_1$, $A_1 A_2$, $A_2 A_3$, etc.

From the property of the circle stated above,

$$\angle T_1 O A_1 = 2 \angle V T_1 A_1 = 2\delta_1$$

Now,

$$R \times 2 \delta_1 = T_1 A_1 = C_1$$

Or $$\delta_1 = \frac{C_1}{2R} \text{ radian}$$

Or $$\delta_1 = \frac{C_1}{2R} \times \frac{180^\circ}{\pi} \text{ degree}$$

Or $$\delta_1 = \frac{C_1}{2R} \times \frac{180^\circ}{\pi} \times 60 \text{ minutes}$$

Or $$\delta_1 = 1718.9 \frac{C_1}{2R} \text{ minutes}$$

Similarly,

$$\delta_2 - 1718.9 \frac{C_2}{2R} \text{ minutes,}$$

$$\delta_3 = 1718.9 \frac{C_3}{2R} \text{ minutes}$$

For first chord $T_1 A_1$,

The deflection angle = Its tangential angle

$$\Delta_1 = \delta_1$$

For second chord A_1A_2,

The deflection angle = $\Delta_2 = \angle VT_1 A_2$

The tangential angle for chord $A_1 A_2 = \delta_2$ = Angle between tangent at A_1 and chord $A_1 A_2$

The angle subtended by the chord $A_1 A_2$ at $T_1 = \angle A_1T_1 A_2 = \delta_2$

Then, $\Delta 2 = \angle VT_1A_2 = \angle VT_1 A_1 + \angle A_1T_1 A_2$

$= \delta_1 + \delta_2$

$= \Delta_1 + \delta_2$

Similarly, $\Delta_3 = \Delta_2 + \delta_3$

And $\Delta_n = \delta_1 + \delta_2 + ...+ \delta_n$

$= \Delta_n - 1 + \delta_n$

The deflection angle of long chord $T_1T_2 = \angle V T_1T_2 = \Delta/2$

Check: Sum of all individual tangential angles is equal to half of the deflection angle of the circular curve.

Procedure to set out the curve on the field:

1. Establish P.C. (T_1), P.T. (T_2) and P.I. (V) in the field.
2. Set the theodolite at T_1.
3. Bisect the P. I. and set the vernier A to read zero, clamp lower plate.
4. Release the upper plate and turn the telescope to read vernier A as Δ_1, hence the line of sight is along $T_1 A_1$.
5. Holding zero end of the tape at T_1, take distance $T_1 A_1 = C_1$, and swing the tape with an arrow till the arrow is bisected by the telescope. Thus point A_1 is established.
6. Now turn the telescope to read vernier A as Δ_2, hence the line of sight is along $T_1 A_2$.
7. Holding zero end of the tape at A_1, take distance $A_1 A_2 = C_2$, and swing the tape with an arrow till the arrow is bisected by the telescope. Thus point A_2 is established.
8. Repeat the process till last point T_2 is established.

UNIT SUMMARY

- **Curve:** A gradual change in direction is accomplished at the intersection by introducing the curves.
- **Types:** There are mainly two types
  - (i) Horizontal Curve
  - (ii) Vertical Curve

    Horizontal curve is further classified as under:
    - (a) simple circular curve
    - (b) compound curve
    - (c) reverse curve
    - (d) transition curve
- **Element of simple circular curve:**

1. Tangent length = R tan Δ /21
2. Length of the long chord = 2RSin Δ /2

3. Length of the curve = $\pi R\ \Delta/180°$
4. Chainage of the point T_1 = Chainage of V – Tangent length
5. Chainage of the point T_2 = Chainage of T_1 + Length of curve

- **Designation of curve:**

  A simple circular curve is designated by following ways:

  (i) By radius of arc

  (ii) By degree of a curve

  For small value of degree of curve, both arc definition and chord definition give same result.

  Hence, $D_a = D_c = \dfrac{1718.88}{R}$ Degree

- **Offsets from the long chord method:**

$$O_x = \sqrt{(R^2 - x^2)} - (R - O_o)$$

- **Deflection angle by Rankine's method:**

$$\delta_1 = 1718.9 \frac{C_1}{2R} \text{ minutes}$$

$$\delta_2 = 1718.9 \frac{C_2}{2R} \text{ minutes}$$

$$\delta_3 = 1718.9 \frac{C_3}{2R} \text{ minutes}$$

$$\Delta_1 = \Delta_2 + \delta_1$$

EXERCISES

Multiple Choice Questions.

1. What is the tangent length of simple circular curve having length of 300 m and angle of deflection is 30°.

(a) 300.00 m (b) 165.67 m

(c) 173.20 m (d) 256.38 m

2. The formula for the tangent length of the simple circular curve is

(a) $R \tan \Delta/2$ (b) $R [\sec \Delta/2 - 1]$

(c) $2R\sin \Delta/2$ (d) $R \sin \Delta/2$

3. The deflection angle of any chord is equal to the deflection for the previous chord plus the tangential angle for that chord.

(a) True (b) False

4. Angular method of laying out of simple circular curve is

(a) Rankine's method of deflection angle (b) Successive bisection of arcs

(c) Offsets from the chord produced (d) Perpendicular offsets from the tangent

5. The shape of simple circular curve is
   (a) Parabolic (b) Spiral
   (c) Circular (d) Elliptical
6. A curve consists of two or more simple arcs that turns in the same direction and join at the common tangent points is known as
   (a) Simple curve (b) Reverse curve
   (c) Compound curve (d) Vertical curve
7. If the angle of intersection of a curve is θ, then the deflection angle will be
   (a) $\theta/2$ (b) $180° - \theta$
   (c) $180° + \theta$ (d) $90° + \theta$
8. if the radius of the curve is 200 m and length of arc is 30 m, then what will be degree of curve by arc definition.
   (a) 2° 24' 21" (b) 7° 38' 32"
   (c) 8° 35' 42" (d) 5° 38' 42"
9. When two tangents PQ and QR meet at Q, the point Q is known as
   (a) Summit (b) Apex
   (c) Vertex (d) Non of these
10. The central angle of the simple circular curve is not equal the deflection angle.
   (a) True (b) False

ANSWERS															
1.	(c)	2.	(a)	3.	(a)	4.	(a)	5.	(c)	6.	(c)	7.	(b)	8.	(c)
9.	(c)	10.	(b)												

ANSWERS TO MULTIPLE CHOICE QUESTIONS

Short Questions:

1. Classify curves.
2. Explain simple circular curve with neat sketch.
3. Explain compound curve with neat sketch.
4. Explain reverse curve with neat sketch.
5. Name the various elements of simple circular curve.
6. Define following:
   (i) P. C.
   (ii) P.I.
   (iii) P.T.
7. How is simple circular curve designated?
8. Derive relationship between degree of curve and radius with arc definition.
9. Name linear methods of setting out simple circular curves.
10. Name angular methods of setting out simple circular curves.

Long Questions:

1. Describe the types of curves
2. Explain various elements of simple circular curve with neat sketch.
3. How is simple circular curve designated? Derive relationship between degree of curve and radius with (i) arc definition (ii) chord definition
4. Describe, offsets from the long chord method, for setting out a simple circular curve
5. Describe, Rankine's method of deflection angle, for setting out a simple circular curve.

Solved Numerical Problem:

Numerical 1: The tangents intersect at a chainage of 1400 m, the deflection being 26°. Calculate the following quantities for setting out a curve of radius 280 m.

(a) Tangent length

(b) Length of the long chord

(c) Length of the curve

(d) Chainage of the point of commencement

Solution: (a) Tangent length = $R \tan \Delta/2 = 280 \times \tan 26°/2 = 64.64$ m

(b) Length of the long chord = $2R\text{Sin } \Delta/2 = 2 \times 280 \times \sin 26°/2 = 125.97$ m

c) Length of the curve = $\pi R\, \Delta/180° = \pi \times 280 \times 26°/ 180° = 127.05$ m

d) Chainage of the point of commencement = 1400 – 64.64 = 1335.36 m

Numerical 2: Calculate the co-ordinate of the simple circular curve having long chord 160 m and mid ordinate 10 m. Ordinates are measured from long chord at 10 m interval.

Solution: Given,

Length of chord (L) = 160 m

Mid ordinate (O_0) = 10 m

To calculate radius of curve,

$$O_0 = R - \sqrt{R^2 - (L/2)^2}$$

$$O_0 = R - \sqrt{R^2 - (160/2)^2}$$

$$10 = R - \sqrt{R^2 - 80^2}$$

$$\sqrt{R2 - 802} = R - 10$$

both side square

$$(\sqrt{R^2 - 80^2})^2 = (R - 10)^2$$

$$R^2 - 80^2 = R^2 + 10^2 - 2 \times R \times 10$$

$$20\,R = 6500$$

$$R = 325 \text{ m}$$

The ordinates are calculated at 10 m interval from the Centre towards tangent point, by using the equation

$$O_x = \sqrt{R^2 - (X)^2} - (R - O_0)$$

$$O_{10} = \sqrt{325^2 - X^2} - (325 - 10)$$

$$O_{10} = \sqrt{325^2 - 10^2} - (325 - 10)$$

$$O_{10} = 9.82 \text{ m}$$

$$O20 = \sqrt{325^2 - 20^2} - (325 - 10)$$
$$= 9.38 \text{ m}$$

$$O_{30} = \sqrt{325^2 - 30^2} - (325 - 10)$$
$$= 8.612 \text{ m}$$

$$O_{40} = \sqrt{325^2 - 40^2} - (325 - 10)$$
$$= 7.52 \text{ m}$$

$$O_{50} = \sqrt{325^2 - 50^2} - (325 - 10)$$
$$= 6.13\text{m}$$

$$O^{60} = \sqrt{325^2 - 60^2} - (325 - 10)$$
$$= 4.41 \text{ m}$$

$$O_{70} = \sqrt{325^2 - 70^2} - (325 - 10)$$
$$= 2.37 \text{ m}$$

$$O_{80} = \sqrt{325^2 - 80^2} - (325 - 10)$$
$$= 0.00 \text{ m}$$

Numerical Problem 3 –

Two straights intersect at a chainage 1300 m. deflection angle is 34°. Calculate the required data for setting out the simple circular curve of radius 220 m, by deflection angle method. The peg interval of 20m chain.

Solution:

Radius of curve (R) = 220 m

Deflection angle (Δ) = 34°

Chainage of intersection point (V) = 1300 m

Peg interval = 20m

1. Tangent length = $T_1V = T_2V$

$$= OT_1 \tan\frac{\Delta}{2} = R\tan\frac{\Delta}{2}$$

$$= 220 \times \tan\frac{34^\circ}{2} = 67.26 \text{ m}$$

2. Length of curve $= \dfrac{2\pi R}{360^\circ} \times \Delta = \dfrac{\pi R\Delta}{180^\circ} = \dfrac{\pi \times 220 \times 34^\circ}{180^\circ} = 130.55\text{m}$

3. Chainage of tangent point

Chainage of T_1 = Chainage of V – Tangent length

Chainage of T_1 = 1300 – 67.26 = 1232.74 m

Chainage of T_2 = Chainage of T_1 + length of curve

Chainage of T_2 = 1232.74 + 130.55 = 1363.29

4. The curve length is divided into initial, full and final sub chord length,

Length of sub chord = C_1 = 1235 – 1232.74 = 2.26

Length of (20m) full chord = C_2 to C_7 = 20 × 6 = 120 m, chainage at 1355 m

Length of final sub chord = 1363.29 – 1355 = 8.29 m

5. Deflection angle for initial sub chord,

$$\delta 1 = \frac{C_2}{2R}\text{radian} = \frac{180°C_1}{2\pi R}\text{Degrees} = \frac{180°\times 60C_1}{2\pi R}\text{Minutes}$$

$$= 1718.9\frac{C_1}{R}\text{Minutes} = 1718.9\frac{C_1}{R\times 60}\text{Degree}$$

$$\delta_1 = 1718.9\times\frac{2.26}{220} = 0°\ 17'\ 39'' = 17°\ 39'$$

$$\delta_2 \text{ to } \delta_7 = 1718.9\times\frac{20}{220} = 156°\ 15'$$

$$\delta_8 = 1718.9\times\frac{8.29}{220} = 64°\ 46'$$

Peg at point	Chainage	Length of chord (C)	Deflection angle (δ)	Total deflection (Δ)
T1 (Starting point)	1232.74	0	0	0
C_1	1235	2.26	0°17' 38"	0°17' 38"
C_2	1255	20	2°36'15"	2°53' 53"
C_3	1275	20	2°36'15"	5°30' 08"
C_4	1295	20	2°36'15"	8°6' 23"
C_5	1315	20	2°36'15"	10°42' 38"
C_6	1335	20	2°36'15"	13°18' 53"
C_7	1355	20	2°36'15"	15°55' 08"
T2 (End point)	1363.29	8.29	1°4'46"	16°59' 54"

Arithmetical Check

Total deflection angle(Δ),

= 0°17'38"+2°36'15"+2°36'15"+2°36'15"+2°36'15"+2°36'15"+2°36'15"+1°4'46"

= 16°59' 54"

Numerical Problem 4: Two straight lines T_1L and T_2L intersect at chainage (380 + 15), the angle of deflection being 115°. calculate the chainage of the tangent points of a right handed circular curve of 500 m radius.

Solution: Chainage of L = (380 + 15) = 380 (chains) + 15 (links)

Deflection angle, Δ = 115°

Tangent length = Rtan115°/2

= 410 tan 57.5

= 643.57m

Length of curve = $\pi R\Delta/180°$

= $\pi \times 410 \times 115°/180°$

= 822.50m

Let us assume that length of chain=20m

Chainage of P.I. =380 (chains)+15 (links) = =7603m

Deduct tangent length =	– 643.57m
Chainage of P.C=	6959.43m
Add length of curve=	+ 822.50m
Chainage of P.T. =	7781.93m

Chainage of P.C. = (347 +97.15)

Chainage of P.T. = (389 +9.65)

Unsolved Numerical Problem

Numerical Problem 1-

The tangents intersect at a chainage of 1450 m. the deflection being 28°. Calculate the following quantities for setting out a curve of radius 285 m.

(a) Tangent length
(b) Length of the long chord
(c) Length of the curve
(d) Chainage of the point of commencement

Solutions:

(a) Tangent length = 71.05 m
(b) Length of the long chord = 137.895 m
(c) Length of the curve = 139.27m
(d) Chainage of the point of commencement = 1378.95 m

Numerical Problem 2 – Two straights intersect at a chainage 1230 m. deflection angle is 28°. Calculate the required data for setting out the simple circular curve of radius 180 m, by deflection angle method. The peg interval of 20 m chain.

Answer:

(a) Tangent length = 44.88 m
b) Length of the curve = 87.96 m
(c) Chainage of T_1 = 1185.12 m, Chainage of T_2 = 1273.08 m
(d) Deflection angle = $\delta_1 = 2° 22' 5''$, δ_2 to $\delta_4 = 3° 10' 59''$, $\delta_5 = 2° 4' 54''$

Numerical Problem 3: Two straight lines T_1L and T_2L intersect at chainage (400 +20), the angle of deflection being 112°. Calculate the chainage of the tangent points of a right handed circular curve of 410m radius.

Answer: Chainage of P.C = (363 + 13.6)

Chainage of P.T. = (412+0.5)

Numerical Problem 4:

Calculate the co-ordinate of the simple circular curve having long chord 170 m and mid ordinate 7m. Ordinates are measured from long chord at 10m interval.

Answer:

$O_{10} = 6.90$ m	$O_{40} = 5.45$ m	$O_{70} = 2.26$ m
$O_{20} = 6.62$ m	$O_{50} = 4.59$ m	$O_{80} = 0.804$ m
$O_{30} = 6.13$ m	$O_{60} = 3.52$ m	$O_{85} = 0.00$ m

PRACTICAL

Objective:

Set out a circular curve by Rankine's Method of Deflection Angles.

Instruments required:

Theodolite, Ranging rods, Chain, Arrows and pegs.

Required data:

Chainage of curve, Deflection angle of curve (Δ) and radius of curve (R)

Procedure:

Prepare a table of deflection angles for the first sub-chord, normal- chords and last sub-chord using formula $\delta_1 =$ 1718.9 minutes. And $\Delta_1 = \delta_1$, $\Delta_2 = \Delta_1 + \delta_1$

Establish P.C. (T_1) , P.T. (T_2) and P.I. (V) in the field.

Set the theodolite at T_1.

Bisect the P. I. and set the vernier A to read zero, clamp lower plate.

Release the upper plate and turn the telescope to read vernier A as Δ_1, hence the line of sight is along $T_1 A_1$

Holding zero end of the tape at T_1, take distance $T_1 A_1 = C_1$, and swing the tape with an arrow till the arrow is bisected by the telescope. Thus point A1 is established

Now turn the telescope to read vernier A as Δ_2, hence the line of sight is along $T_1 A_2$

Holding zero end of the tape at A_1, take distance $A_1 A_2 = C_2$, and swing the tape with an arrow till the arrow is bisected by the telescope. Thus point A_2 is established.

Repeat the process till last point T_2 is established.

Result: The given simple curve is thus set-out.

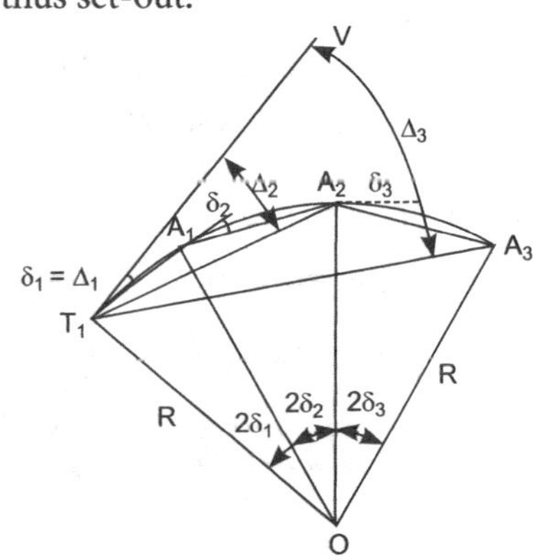

Fig. 4.9: Setting out Circular Curve by Rankine's Method

KNOW MORE

Other types of curves in surveying

1. **Combined Curve**: The combination of a transition curve and a simple circular curve is called as a combined curve.
2. **Summit curve**: A vertical curve with its convexity in the upwards direction is called as a summit curve
3. **Valley curve**: A vertical curve with its convexity in the downwards direction is called as a valley curve. It is also named as sag curve.

Obstacles to the location of curves

Case 1: when the P.I. is inaccessible

Case 2: when the P.C. is inaccessible

Case 3: when the P.T. is inaccessible

Case 4: when both P.C. and P.T. are inaccessible

Case 5: when both P.C. and P.I. are inaccessible

Case 6: when complete curve cannot be set-out from P.C.

Case 7: when obstacles in chaining

The procedure of setting out the curve with above obstacles can be known in further study.

REFERENCES AND SUGGESTED READINGS

1. B.C Punmia, Surveying Vol –II, Laxmi Publication (P) Ltd, 2023.
2. N.N. Basak, surveying & Levelling,McGraw hill Iindia, Private Limited, Noida, 2017.
3. Bhavikatti SS, Surveying and Levelling Vol. II, I.K International Publishing House Pvt. Ltd, 2016.
4. R. Agor, A Text Book of Surveying & Levelling, Khanna Publication, 2015.
5. S.K. Duggal, Surveying Vol. I, McGraw Hill Publishing Company Ltd., 2013.
6. C. Venkatramaiah, Textbook of Surveying, University Press (India) Limited, 2011.
7. Saikia MD, Das BM, Das MM, Surveying, Prentice Hall India Learning Limited, 2010.
8. T.P. Kanetkar and S.V. Kulkarini, Surveying and Levelling, Vidyarthi Griha Prakashan, 1967.
9. https://www.ikbooks.com/home/samplechapter?filename=121_Sample_Chapter.pdf
10. https://priodeep.weebly.com/uploads/6/5/4/9/65495087/_simple_curves_updated.pdf

Dynamic QR Code for Further Reading

Simple Circular Curve Notes

Fundamentals of surveying book by S. K. Roy

Simple circular curve by Rankine's method of def...

Elements of simple circular curves

Video lecture by NPTEL

5 Advanced Surveying Equipment

UNIT SPECIFICS

Through this unit we have discussed the following aspects:

- Principle of Electronic Distance Meter (EDM)
- EDM component parts and their functions
- Use of EDM
- Use of micro-optic Theodolite and Electronic Digital Theodolite
- Use of Total Station and use of its function keys
- Measurements of horizontal angles, vertical angles, distances, and coordinates using Total Station
- Traversing, Profile Survey and Contouring with Total Station

RATIONALE

This unit explains the principle of EDM. The EDM component parts and their functions are described. The use of EDM is discussed. The use of micro optic Theodolite and Electronic Digital Theodolite is elaborated. The utility of Total station and use of its function keys is discussed. The measurement of horizontal and vertical angles, determination of distances and coordinates using Total Station is explained. The finding of field data through traversing, profile survey and contouring using Total station is presented.

PRE-REQUISITE

Mathematics: Basic knowledge of Trigonometry

Basic Surveying (Third Semester Diploma)

UNIT OUTCOMES

List of outcomes of this unit is as follows:

U5-O1: State the principle of EDM

U5-O2: Describe EDM components, its functions and use of EDM

U5-O3: Make use of micro-optic Theodolite and Electronic Digital Theodolite

U5-O4: Explain Total Station and use of its function keys

U5-O5: Determine field data using Total Station

Unit-5 Outcomes	EXPECTED MAPPING WITH COURSE OUTCOMES (1- Weak Correlation; 2- Medium correlation; 3- Strong Correlation)				
	CO-1	CO-2	CO-3	CO-4	CO-5
U5-O1	-	1	1	2	1
U5-O2	-	1	1	2	1
U5-O3	1	3	3	2	1
U5-O4	1	2	2	3	1
U5-O5	1	2	2	3	1

5.1 INTRODUCTION

As discussed in earlier chapter, When the ground is rough and the work is to be performed speedily with ordinary precision, tacheometry may be a good option and sometimes preferred over chaining. But in tacheometry, range is limited and accuracy is moderate.

When high degree of precision is required and larger range is needed, EDM may be preferred. EDM works on the properties of electromagnetic waves

In this chapter, the use of micro-optic Theodolite and Electronic Digital Theodolite is also illustrated. The Total Station, which is combination of Digital Theodolite and EDM, is elaborated.

5.2 PRINCIPLE OF EDM

The EDM is based on the properties of electromagnetic waves. As per the basic wave theory, electromagnetic energy travels in harmonic sinusoidal fashion at the velocity of light (c) 3×10^8 m/s. The distance between two successive crests or troughs is the wavelength (λ). Number of peaks passing through a fixed point per unit time is wave frequency (υ). Electromagnetic wave is shown in Fig. 5.1.

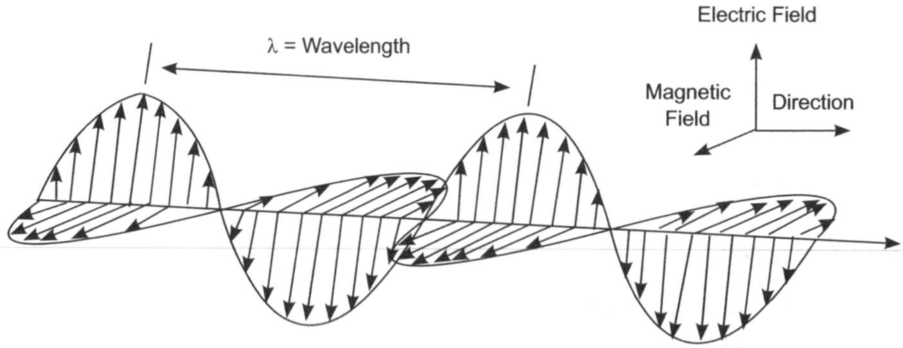

(a)

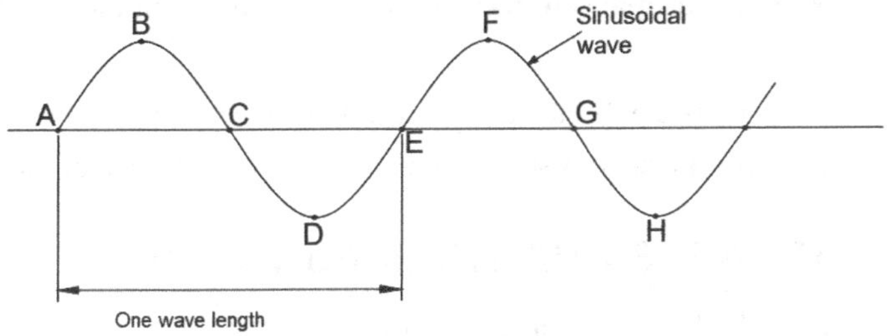

(b)

Fig. 5.1: Electromagnetic wave in harmonic sinusoidal fashion

Phase of the wave: The wave may be at different amplitude at particular point of time, this is called phase of the wave (ϕ). The phase value at different points (Fig. 5.1) of the wave shown in the Table 1.

Table 5.1

Point	A	B	C	D	E	F	G	H
Phase (ϕ)	0	90	180	270	360	90	180	270

The type of wave is distinguished by its wavelength or its frequency. Continuous spectrum of Electromagnetic energy is classified into various waves based on its wavelength and frequency for simplicity. Wave length of electromagnetic energy ranges from cosmic rays (Short waves) to radio waves (long waves), in between the extremities the other wave length regions are ɣ-rays, X-rays, UV-rays, Visible region, Infrared region, Microwave region.

Measurement of the length AB (Fig. 5.2): Let length of the survey line AB (L) is to be measured. A transmitter is placed at A, which emits waves. And if a receiver is placed at B, which receives the wave, it is very difficult to start a timer at B when a wave is transmitted at A as transmit time is of order of 1 x 10-6 sec. Hence a reflector is kept at B instead of a receiver. The reflector reflects the waves back and they are received at A. Thus, the transmitter at A also acts as receiver. The wave travels double distance that of length AB (L). Hence in the Fig. 5.2.

$$2L = n\lambda + \Delta\lambda \qquad ...(5.1)$$

Where,

L = distance between A & B

λ = wavelength

n = whole number of wavelengths moved by the wave

$\Delta\lambda$ = fraction of the wavelength moved by the wave

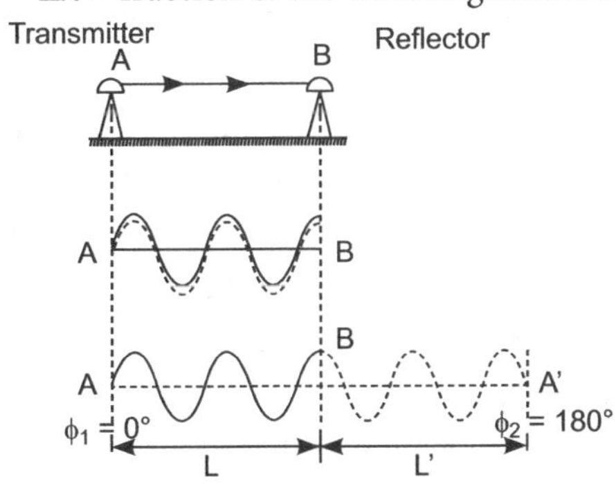

Fig. 5.2: Distance Measurement by EDM

$\Delta\lambda$ is determined by phase comparison techniques. Modern techniques can measure 1/1000 part of a wavelength.

Modulation: In order to increase the accuracy, modulation technique is used where the measuring wave which is used for phase comparison is superimposed on a carrier wave of more higher frequency .

5.3 COMPONENT PARTS AND THEIR FUNCTION

The basic components of an EDM are as under: (Fig. 5.2)

1. Transmitter – The transmitter propagates the electromagnetic waves and receives the wave after reflection. It is placed at starting point of the line
2. Reflector - The reflector reflects the wave. It is placed at end point of the line.

5.4 TYPES AND USES OF EDM INSTRUMENTS

Based on carrier wave, EDM instruments may be classified as under:

(i) Microwave instruments

(ii) Visible light instruments

(iii) Infrared instruments

All three types instruments use short wavelength or higher frequency.

In Microwave instruments, carrier wave frequency is of range of 3 to 30 GHz, can measure distance up to 100 km. The instrument Tellurometer falls under this category.

In Visible light instruments, carrier wave frequency is of order of 5 x 1014 Hz, can measure distance up to 2 to 3 km. The instrument Geodimeter falls under this category.

In Infrared instruments, carrier wave wavelength is about 0.9 μm, can measure distance up to 2 to 5 km. The instrument Wild Distomats falls under this category.

5.5 MICRO-OPTIC THEODOLITE

The micro-optic theodolite is also known optical or glass arc theodolite. In this type of theodolite, graduated circles are made of glass. The graduation is etched on glass disc photographically. The reading is reflected into the microscope by a system of prisms and lenses. The eyepiece of telescope and the eyepiece of the microscope lie very near to each other; hence the observer can see the reading and the object simultaneously without changing his position. The system of prisms and mirrors are so arranged that the readings at diametrically opposite ends on circle are taken by the microscope simultaneously. At the same time the readings on horizontal and vertical circles can be taken by the single microscope.

The graduations of the circle are illuminated by the light beam, hence smaller reading can be read by the microscope. Hence the least count of both horizontal and vertical circle of many theodolites is kept 20" and vernier scale is divided in 20 parts, hence the least count of the instruments is 1".

The micro-optic theodolite is used for fast and precise measurements. It is used in triangulation, precise traversing, cadastral survey, tunnelling, and variety of tasks.

Fig. 5.3: Photograph of Micro-Optic Theodolite

5.6 ELECTRONIC DIGITAL THEODOLITE

An Electronic theodolite is light and compact instrument. When it is used with EDM, it becomes Total Station. The basic principle of it, is like that of a vernier theodolite. The system of Electronic digital theodolite resolves angular values and display them in digital form. The display system uses light emitting diodes (LEDs) or liquid crystal diodes (LCDs). The glass circle is used in Electronic theodolite and it is coded in special manner. The light is passed through the encoded circle and the patterns of the emerging light is detected by photodiodes. The rotation of the circle varies with light intensity. The variation in light intensity is related to angle of rotation and converted into electrical signals by the photodiodes and these electrical signals are sent to microprocessor which converts them into angular output. This is the principle of Electronic digital theodolite.

The electronic theodolite is used for following reasons

1. They provide greater accuracy
2. The readings are in digital form, convenient to note quickly
3. 1The horizontal circle can be instantly put at zero or set to any other value.
4. The horizontal circle readings can be taken either to the right or left of zero.
5. The method of repetition is not needed.

Fig. 5.4: Photograph of Electronic Theodolite

5.7 TOTAL STATION

Total station is an advanced surveying instrument which is combination of Electromagnetic Distance Meter and Electronic theodolite. The microprocessor, electronic data collector and storage system are also integrated in the Total station. The instrument is used to measure horizontal & vertical angles and sloping distance of object to the instrument. The prism reflector mounted on pole is placed at the object point. The laser from the EDM bisects the prism reflector. Some total station works without reflector.

Microprocessor of a Total station can process the data to determine:

1. Mean of number of angles measured.
2. Mean of number of distance measured.
3. Horizontal distance.
4. Distance between two stations
5. Elevation of objects
6. Coordinates of the points.

The values measured and saved in a Total Station may be transferred to computers for required use

5.7.1 Fundamental Quantities Measured by Total Station

(i) **Horizontal angle** - The angle between optical axis of the instrument and the instrument north in horizontal plane called horizontal angle.

(ii) **Vertical angle** - The inclination of the optical axis from the vertical is called vertical angle

(iii) **Slope distance** - The distance between the target and the instrument is called slope distance

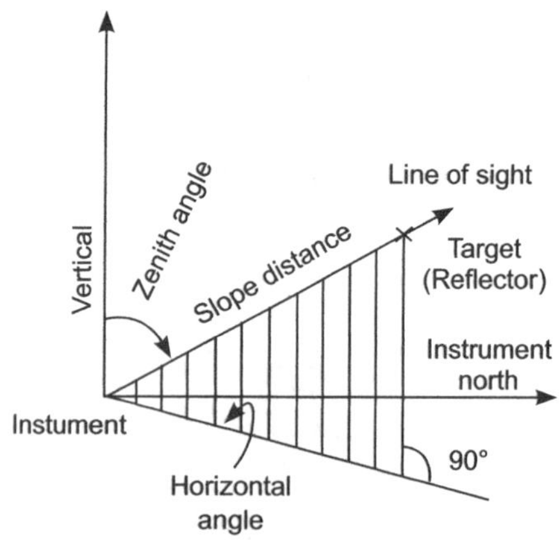

Fig. 5.5: Basic Measurements by Total Station

Measurement of Horizontal angle:

Horizontal angle is measured from the assumed north. This direction is set as zero direction The user choose a direction of his choice as zero direction. This direction may be towards some prominent feature or may be approximate true or magnetic north.

Measurement of Vertical Angle :

Measuring vertical angle needs that the instrument must be perfectly vertical. Total station has an internal sensor that can detect small deviations of the instrument from the vertical. If it is too far from the vertical the instrument will provide tilt error message. The vertical angle is measured from true vertical line.

Measurement of Slope Distance:

The major part of the Total station is EDM. It always measures the sloping distance from the instrument station to the object. The range of slope distance that can be measured from Total station varies from 3 km to 5 km.

5.7.2 Control Panel of Total Station

Various brands' Total station has different types of control panels. They are more or less similar functions.

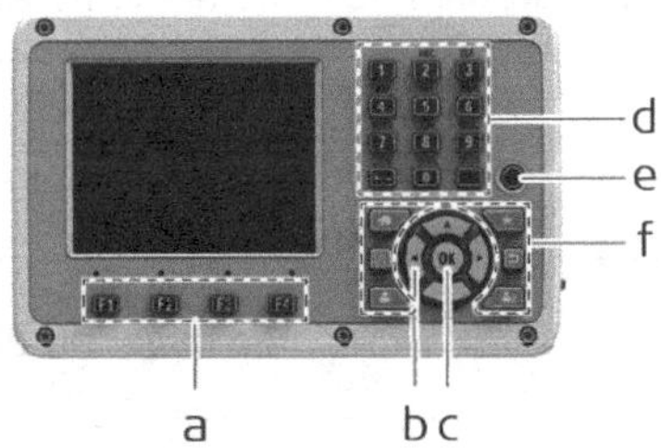

a Function keys **F1** to **F4**
b Navigation keys
c **ENTER** key
d Alphanumeric keypad
e **ON/OFF** key
f Fixed keys

Fig. 5.6: Control Panel of Total Station

Control panel of Leica TS-03 model is shown (Fig. 5.6) It has a LED screen and a key board. Softkeys are selected using the relevant **F1** to **F4** function key. Here the functionality of the common softkeys used by the system, are described below.

5.7.3 Common softkey functions

Table 5.2

S.No.	Key	Function
1	**Cont**	If entry screen: Confirms measured or entered values and continues the process.
		If message screen: Confirms message and continues with selected action or returns to the previous screen to reselect
2	**Back**	To return to the last active screen.

3	**Default**	To reset all editable fields to their default values.
4	**Dist**	To start distance and angle measurements without saving the measured values.
5	**EDM**	To view and change EDM settings.
6	**New**	To open the manual coordinate entry screen.
7	**Find**	To search for an entered point
8	**List**	To display the list of available points.
9	**Meas**	To start distance and angle measurements and save the measured values.
10	**Exit**	To exit the screen or app.
11	**Store**	To save the displayed values.
12	**View**	To display the coordinate and job details of the selected point.
13	**-> ABC**	To change the keypad operation to numerical.
14	↓	To display the next softkey level.
15	↑	To return to the first softkey level.

5.7.4 Orientation and Location of the Instrument

The instrument is set up at the given station. It is centered and levelled. (The centering and levelling are almost similar that of Theodolite). The station setup window is opened. The station name (point ID) and code are fed. If the coordinate of the station is given, then they are fed. One reference point coordinates must also be given, by taking backsight to this reference point the orientation of the instrument is done and exact location of the station is known. If the coordinate of the station is not given, the zero direction is set towards assumed north/convenient direction and the assumed convenient coordinates of the station and height of the instrument are fed. The height of reflector is also fed.

5.7.5 Sighting of the Object and Display of the Result

The prism reflector at object point is sighted by EDM. The Survey window is opened. The screen key 'Meas' is pressed. The results on the screen are displayed as under

1.	**Target Height**	Prism height.
2.	**Code**	Object code
3.	**Hz angle**	Horizontal angle
4.	**V angle**	Vertical angle
5.	**Hor. dist.**	Horizontal distance
6.	**Slope dist.**	Measured slope distance
7.	**Height diff.**	Height difference between station and reflector.
8.	**Easting**	Easting coordinate of measured point.
9.	**Northing**	Northing coordinate of measured point.
10.	**Height**	Height coordinate of the measured point.

Above results provide horizontal angle, vertical angle, horizontal distance and coordinates of object point.

5.7.6 Traversing by Total Station

Total station traversing is similar that of theodolite traversing except the distance between two stations are measured by EDM. The traverse stations are fixed. The instrument is set up at first station. The horizontal angle of previous line and next line are taken. The difference between two horizontal angles will give included angle. Similarly, the instrument is set up at other stations and included angles are determined and traverse is completed. The traverse is created by transferring the data in Auto Cad through Map option.

5.7.7 Contouring by Total Station

The contouring is done by the method of cross sectioning or creating grid. In cross sectioning method, the cross sections are marked at fixed interval on the land. The points are identified at each cross section and each point is numbered. The Total station is set at a convenient point and its coordinates is fed. The prism reflector is placed at every point and sighted by EDM. The elevation coordinates of each point are saved. The map showing cross sections and elevation of each point is created by transferring the data in Auto Cad through Map option. The contours of different elevations are drawn by joining the points of same elevation through interpolation.

Similarly, in grid method, a grid is marked on the land and each grid point is numbered. The contouring is done by the procedure mentioned above.

UNIT SUMMARY

- **Electromagnetic Distance Measurement (EDM)** - The EDM is based on the properties of electromagnetic waves. The distance measurement is based on the following formula.

$$2L = n\lambda + \Delta\lambda$$

Where, L = distance between A & B

λ = wavelength

n = whole number of wavelengths moved by the wave

$\Delta\lambda$ = fraction of the wavelength moved by the wave

- Types of EDM Instruments

  Based on carrier wave, EDM instruments may be classified as under:

  (i) Microwave instruments

  (ii) Visible light instruments

  (iii) Infrared instruments

- **Micro-Optic Theodolite** -The micro-optic theodolite is also known optical or glass arc theodolite. In this type of theodolite graduated circles are made of glass. The micro-optic theodolite is used for fast and precise measurements. It is used in triangulation, precise traversing, cadastral survey, tunnelling, and variety of tasks.
- **Electronic Digital Theodolite** - A Electronic theodolite is light and compact instrument. The basic principle of it, is similar to that of a vernier theodolite. The system of Electronic digital theodolite resolves angular values and display them in digital form.
- **Total Station** - Total station is an advanced surveying instrument which is combination of Electromagnetic Distance Meter and Electronic theodolite. The microprocessor, electronic data collector and storage system are also integrated in the Total station. The instrument is used to measure horizontal & vertical angles and sloping distance of object to the instrument.

EXERCISES

Multiple Choice Questions.

1. In the total station, data is stored in ________.

(a) External hardware (b) Memory card
(c) Micro processor (d) Pen drive

2. Which of the following correct combination of total station?

(a) EDM, GPS (b) Theodolite, EDM
(c) Electronic theodolite, EDM (d) Theodolite, compass

3. What is the advantage of using EDM?

(a) Precise measurement of distance (b) Electronic batteries
(c) Expensive (d) Accuracy affected by atmospheric condition

4. Which of the following is the latest development in a total station

(a) High resolution (b) Automatic
(c) Robotic (d) High accuracy

5. Which of the following is used up to a range of 100 km.

(a) Infrared (b) Visible range
(c) Microwave (d) Ultra-Violet

6. Which of the following are type of EDM instrument?

(a) Microwave (b) Infrared
(c) Visible light (d) All of the above

7. Total station can be used for ________.

(a) Angular measurement (b) Linear measurement
(c) Elevation measurement (d) All of the above

8. In the optical microscope, the graduated circle is made of

(a) Glass (b) Steel
(c) Plastic (d) Aluminium

9. In digital theodolite the results are obtained

(a) In digital form (b) In angular form
(c) Digital and angular form (d) None of above

10. In the theodolite the vertical angle is measured from the plumb line

(a) True (b) False

ANSWERS													
1.	(c)	2.	(c)	3.	(a)	4.	(c)	5.	(c)	6.	(d)	7.	(d)
8.	(a)	9.	(a)	10.	(a)								

Answer To the Multiple Questions

Short Questions:

1. What is principle of EDM?
2. Write types of EDM.
3. When EDM is preferred over Tacheometer?
4. Write uses of EDM.
5. What is micro-optic theodolite?
6. Write advantages Digital theodolite.
7. Write uses of Total Station.
8. Explain function keys of Total station and their uses.
9. What is vertical and horizontal angles in Total station?
10. How contouring is done by Total station?

Long Questions:

1. Describe principle, types, and uses of EDM
2. Explain working and uses of micro-optic theodolite and Digital Theodolite.
3. Describe uses of Total station and uses of function keys of Total station.
4. Explain following parameters and their determination by Total station
   (i) Horizontal angle
   (ii) Vertical angle
   (iii) Horizontal distance
   (iv) Coordinates of a point

PRACTICAL

Experiment No. 1

Objective: Use EDM to measure horizontal distance AB.

Required Equipment: EDM, Tripod, Prism reflector

Procedure:

- Set the EDM on tripod at station A and Switch on.
- Centre and level the instrument.
- Install the reflector on station B.
- Go to the distance Icon by using the Menu bar.
- Target the point B and focus the prism reflector
- Click on "Read", distance measured will be displayed on the screen.
- Record the horizontal distance between two points.

Results:

The horizontal distance between two points A & B is _________.

Experiment No. 2

Objective: Use Total station instrument to measure horizontal angle, vertical angle & horizontal distance.

Required Equipment: Total Station (Lieca TS-03), Tripod, Prism reflector

Procedure:

- Establish the instrument at given station A
- Centre and level the instrument.
- Open the station setup window.
- Feed the height of instrument and height of reflector.
- Feed the name (point ID) and code of the station A.
- Sight the EDM of the instrument towards assumed north direction.
- Set the horizontal angle at zero reading.
- Sight the EDM towards the given object/station B.
- The Survey window is opened.
- Press the screen key "Meas".
- The results on the screen are displayed.
- Record and save the horizontal angle, vertical angle and horizontal distance.

Results:

1. The horizontal distance between two points A & B is _________.
2. The horizontal angle between assumed north and line AB _________.
3. The vertical angle between plumb line (vertical line) and line AB _________.

Experiment No. 3

Objective: Use Total station instrument to carry out Survey Project for closed traverse for minimum five sides.

Required Equipment: Total Station (Lieca TS-03), Tripod, Prism reflector

Procedure:

- Identify the survey stations A, B, C, D and E.
- Establish the instrument at given station A
- Centre and level the instrument.
- Open the station setup window.
- Feed the height of instrument and height of reflector.
- Feed the name (point ID) and code of the station A.
- Sight the EDM of the instrument towards assumed north direction.
- Set the horizontal angle at zero reading.
- Sight the EDM towards the reflector held at station B.
- The Survey window is opened.
- Press the screen key "Meas".
- Record the horizontal angle and horizontal distance AB.
- Sight the EDM towards the reflector held at station E.
- Record the horizontal angle and distance AE.
- The difference between the two horizontal angles will provide the included angle at A.
- Shift the instrument at station at B. and repeat the process and determine the included angles at B, C, D, and E. and complete the closed traverse ABCDEA.
- Transfer the data to Auto-CAD. and by using the map option create the traverse.

KNOW MORE

- **History and principle of Tellurometer:** https://www.fig.net/pub/fig2008/papers/hs01/hs01_03_sturman_wright_2833.pdf
- **Errors in Total Station:** https://youtu.be/GkGf9_8tPsU
- **Introduction to robotic total station:** https://youtu.be/nr6jrvjPH48
- **Distomat:** https://latestnews.fresherslive.com/articles/what-purpose-is-distomat-used-get-to-know-for-what-purpose-distomat-is-used-253182
- **Geodimeter:** https://civillearners.com/2021/07/27/geodimeter-detailed-explanation/

REFERENCES AND SUGGESTED READINGS

1. B.C. Punmia, Surveying – II, Surveying, Laxmi Publications (P) Ltd, 2023.
2. Dr. Narayan R. Chandak, Prof. Hemraj R. Kumavat, Surveying Fundamentals and Advanced Practices, KATSON books, 2019.
3. B.C. Punmia, Surveying – III Higher Surveying, Laxmi Publications (P) Ltd, 2017.

4. Sateesh Gopi, R. Satikumar, N Madhu- Advance surveying, Pearson Education India, 2017.
5. C.L. Kochher, A Text book of surveying, Dhanpat Rai Publication, 2013.
6. https://www.digimat.in/nptel/courses/video/105107158/L17.html
7. https://www.digimat.in/nptel/courses/video/105107158/L18.html
9. https://www.digimat.in/nptel/courses/video/105107158/L38.html
10. https://www.youtube.com/watch?v=bbs5AEPstl4
11. http://www.nitttrc.edu.in/nptel/courses/video/105107158/L25.html

Dynamic QR Code for Further Reading

Following QR codes are given for further study of EDM and Total Station.

Lecture on EDM

Introduction to Total Station

Parts of Total Station

Accessories Used in Total Station

Measurement of Distance by Total Station

Measurement of vertical angle and height by TS

6 Remote Sensing, GPS and GIS

UNIT SPECIFICS

Through this unit we have discussed the following aspects:

- Principle of Remote Sensing System
- Application of Remote Sensing in Civil Engineering
- Use of GPS Instruments
- Overview of GIS and its components
- Applications of GIS and Software used in GIS
- Introduction of Drone Surveying

RATIONALE

This unit explains the principle of remote sensing system. The applications of remote sensing in Civil Engineering are elaborated. The land use, land cover and disaster management through remote sensing are explained. The use of Global Positioning System (GPS) instruments is discussed. The overview of Geographical Information System (GIS) is presented. The components of GIS and application of GIS are described. The software used in GIS are discussed. The introduction of Drone surveying is presented.

PRE-REQUISITE

Mathematics: Basic knowledge of Trigonometry

Basic Surveying (Third Semester Diploma)

UNIT OUTCOMES

List of outcomes of this unit is as follows:

U6-O1: State the principle of remote sensing

U6-O2: Describe the applications of remote sensing in Civil Engineering

U6-O3: Make use of GPS instruments

U6-O4: Explain GIS and its applications

U6-O5: Illustrate Drone Surveying

Unit-6 Outcomes	EXPECTED MAPPING WITH COURSE OUTCOMES (1- Weak Correlation; 2- Medium correlation; 3- Strong Correlation)				
	CO-1	CO-2	CO-3	CO-4	CO-5
U6-O1	1	1	1	1	2
U6-O2	1	1	1	1	2
U6-O3	1	1	1	1	3
U6-O4	1	1	1	1	2
U6-O5	1	1	1	1	2

6.1 INTRODUCTION

Remote sensing is a technique of obtaining information about a target by collecting data with the help of a device that is not in contact with the target. The target may be an object, area or any phenomenon.

A human eye is the best example of remote sensing. The eyes remotely collect the data about an object through sight and send to the mind which in turn interpret the data and provide information about the object. But in the strict sense, the remote sensing makes use electromagnetic energy as means of detecting and obtaining target characteristics.

The Geographical Information System (GIS) is a system which analyses geographical information obtained by remote sensing, through computer hardware and software.

The Global Positional System (GPS) deals with the techniques which determine global position of an object or a feature using data obtained through remote sensing.

There is a close relationship among remote sensing, GPS and GIS.

6.2 REMOTE SENSING: OVERVIEW

In remote sensing, the electromagnetic radiations strike on the target whose characteristics are to be found out. The reflected, scattered, or emitted EM radiations from the target, are sensed by the sensers. The data received by the sensors is analysed to determine the characteristics of the target. The sensors are located on aircrafts or satellites.

Passive remote sensing: It takes the electromagnetic radiations from the sun and measure the reflected/emitted radiations from the target.

Active remote sensing: It takes the electromagnetic radiations from its own source and measure the reflected/emitted radiations from the target.

The reflected or emitted radiations from the targets/ground feature have radiations of different wavelengths/frequencies which form electromagnetic spectrum. (Fig. 6.1). The user makes use of this EM spectrum for target/ ground features characteristics identification.

The representation of data from sensing system may be pictorial or digital that may be interpreted visually or digitally.

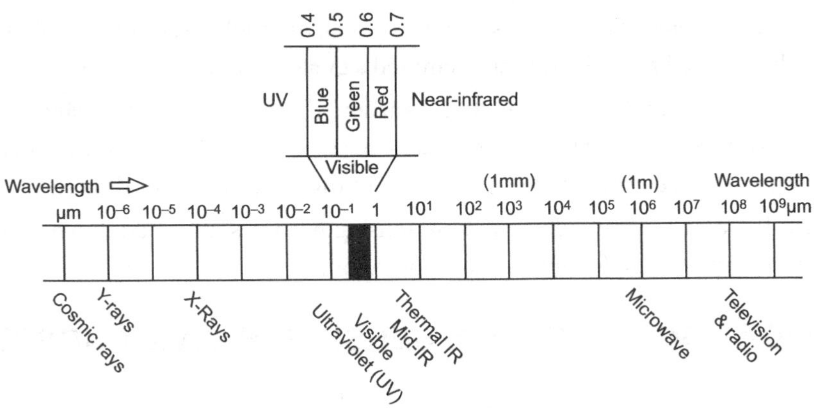

Fig. 6.1: Electromagnetic Spectrum

6.3 REMOTE SENSING SYSTEM

The user should keep following system in mind while adopting remote sensing, this is called a remote sensing system.

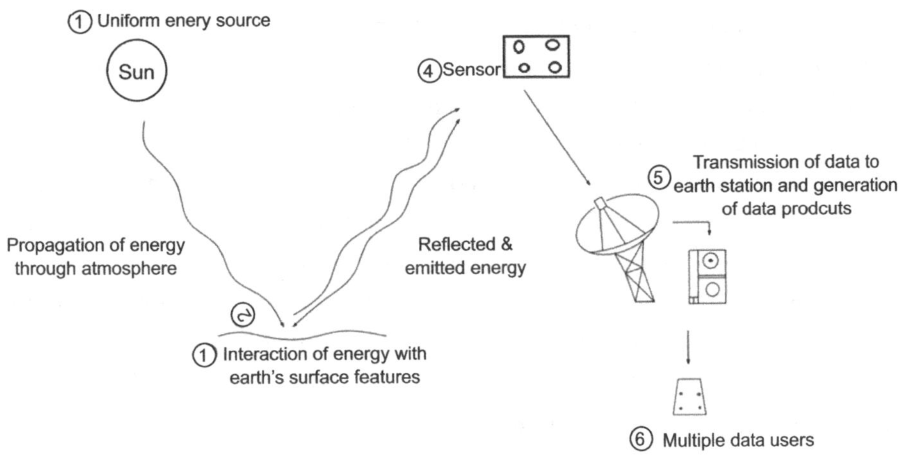

Fig. 6.2: Remote Sensing System

1. **The energy source:** All passive RS system rely on energy which originates from sources other than the sensor itself. It may be either emitted radiation from target/ground features or reflected radiation from the sun. The active RS systems make use of their own source of energy.
2. **Atmosphere:** The electromagnetic radiation in its way from the source to target/ground features and after reflection from target/ground features to sensers , it passes through the atmosphere. The atmosphere modifies the wavelength of the radiation and changes the distribution in EM spectrum.
3. **The energy -matter interactions with the earth's surface:** The interaction of earth surface with EM radiation generates reflected/emitted radiations. The spectrum response patterns of these radiations play significant role in identification and analysis of earth surface features/materials.

4. **The sensors:** The sensors are EM instruments which are highly sensitive to all yielding data and wavelengths received from the reflected/emitted signals . The sensing is a function of wavelength.
5. **The data processing and data product generation:** The data from the sensing system is received at the ground station. The real time data processing system consists of sophisticated recording and visual devices for speedy generation of pictorial or digital product for display.
6. **The users:** The data generated from remote sensing becomes information when someone understands and interpret them for any useful application.

6.4 APPLICATION OF REMOTE SENSING IN CIVIL ENGINEERING

6.4.1 Land use/ land cover mapping

The land use and land cover is very important for planning, development and management activities and is considered as essential parameter for modelling and understanding the earth system. The land cover maps are being developed at local, national and globe levels. The satellite images through remote sensing and arial photographs are being used for land use/land cover mapping.

The term land cover is used to relate the type of features present on the earth surface. Corn field , forest, hills, lakes and highways are some examples of land cover.

The term land use relates to the economic or human activity associated with piece of a land. For example a track of land at outer border of an urban area may be used for either single family housing or for commercial complex. Depending upon the level of mapping detail, its land use may be described as urban use, residential use, or single-family residential use. As per socioeconomic aspects, planning is done for land use, for example a particular land may be used for school or municipal corporation office or income tax office as per planning and importance of the land.

As per hydrological study of rainfall-runoff characteristics, it is important to know the percentage distribution of area for roofs, pavements, grass and trees in a particular track.

The widespread use of remote sensing in land use/ land cover mapping is due to following reason:

(a) Images of very big areas can be acquired quickly.

(b) The problem of access, that prevails in ground surveying, is eliminated.

(c) The RS images provide perspective view, that is difficult in ground survey.

(d) Images interpretation is rapid and cheaper than ground survey.

6.4.2 Disaster Management

Natural disasters and man-made disasters cause loss of life and damage of property. Remote sensing can not prevent the disaster but can be used to minimise the damage due to disaster by following ways:

(a) By identifying and analysing geological formations available in the area and hence detecting hazard prone areas.

(b) By predicting forthcoming hazards based on the ground movements identified by remote sensing.

(c) By assessing real damage occurred and hence conducting proper relief and rescue operation .

6.5 GPS INSTRUMENTS

Global Positioning System (GPS) - It is a radio navigation system used on earth, water bodies, and in air to determine the absolute location, time and velocity of an object irrespective of climate/weather condition.

6.5.1 Elements of GPS

GPS consists of following parts

(i) Satellites

(ii) Ground Stations

(iii) Receivers

6.5.2 Use of GPS Instruments

Satellites act like stars and they send out signals by which their absolute positions are known.

The ground stations are equipped with radars to find out the exact positions of the satellites.

A receiver is a device that may be fitted in mobile phones or in moving objects like cars which continuously receives the signals from the satellites. The receiver can calculate its distance from four or more satellites, which defines absolute position of the receiver.

6.5.3 Principle of GPS

The satellites send the position and time information to the receiver at fixed intervals. The signals are sent to the receiver by radio waves. The time difference between the signal sent from satellite to the signal received by the receiver determines the distance between the receiver and the satellite. The receiver locates its position by obtaining the signals from at least three satellites.

To determine 2-D location, means latitude and longitude, at least 03 satellites are needed. For a 3-D location, means latitude, longitude, and altitude, at least 0 4 satellites are required.

6.6 GEOGRAPHICAL INFORMATION SYSTEM (GIS)

Geographic Information System (GIS) is a system which combines hardware , software, geographical data and human resources data in a smart way to capture, store ,evaluate, update ,manipulate and analyse the data and display them for building effective management system for society or for an organisation.

Through surveying, photogrammetry and remote sensing, data are generated. Using hardware and software, the data are analysed and displayed in form of tables, graphs, and maps. With the help of these types of output using data management system, interpretations and conclusions are drawn and solutions are suggested.

6.6.1 Components of GIS – GIS consists of following components

(i) Hardware: It is used to store and process the data and display results. It includes processor, hard disk, memory unit, display screen, scanner, printer, and plotter etc.

(ii) Software: Software are used to perform various operations needed in GIS to analyse and modify the data.

(iii) Man power: They are experts who develop and run the software and exercise various tasks and processes used in GIS

(iv) Information: It includes geographical information and descriptions about various features.

6.6.2 Application of GIS

GIS is used in various field. Some major applications are mentioned below:

(i) **Agriculture:** Following are the applications in the field of agriculture–

  (a) Drought: Recognising the drought affected area to prevent further damage of seeds, fertilizers and labours.

  (b) Pest control: Predicting the insect and pest attacks allowing farmers and planners to plan accordingly.

  (c) Soil Analysis: Soil analysis may be performed on the basis of previous data sets and satellite images.

  (d) Planning of food demand: As per population's demand, farmers and planners can plan out necessity of production of required crops to avoid hassle.

(ii) **Urban and Town Planning:** Following are the applications in the field of urban and town planning-

  (a) To plan futuristic township

  (b) It becomes easier for planners and architects to hardly miss out on any piece of land

  (c) To plan mega projects

  (d) To tackle water clogging at the time of flood

  (e) To handle cloud burst in monsoon

  (f) To design drainage models

(iii) **Oil Spill:** GIS is used to curb and handle havoc spreading spills in the water bodies.

(iv) **Disaster Management:** GIS is used in disaster management to store the information related to disaster-prone areas like hospitals and isolation camps near risky zones, habitation hub, history of the calamities in the area and effect of previous disasters to prepare contingency plans.

(v) **Navigation and Mapping:** Navigation has become very easy through GIS. Google Maps are widely used for navigation purpose. People use Google maps while travelling to search new places and to locate details. Cab and taxi services work on Google maps.

(vi) **Dam and Reservoir Site location:** GIS is used to locate the perfect site for dam or reservoir construction. GIS considers various factors like engineering factors, environmental conditions, and economic factors before choosing the site.

(vii) **Vegetation and Deforestation Management:** GIS is used to track deforestation and hence subsequently analyse plans for reforestation and vegetation based on the past information of different periods.

(viii) **Business and Marketing Strategies:** GIS is used to store the data of target customers, sales territories, and marketing campaigns for promotion of business and sales.

6.6.3 GIS software Following software are mainly used in GIS

(i) ARC/INFO: This package is used for automation, manipulation, analysis, and digital display of geographical data.

(ii) PAMAP: This GIS package is used for data capturing , storage and analysis.

(iii) GENAMAP: It is versatile software, handles all types of data

(iv) SPANS: It is powerful modelling software used for various applications.

(v) INTERGRAPH: It provides Modular GIS Environment (MGE) for mapping for infrastructure and natural and environmental management.

(vi) ISROGIS: It was developed by Indian Space Research Organisation (ISRO), used in space activities.

6.7 INTRODUCTION TO DRONE SURVEYING

In a drone surveying, a drone is used, it is nothing but an unmanned aerial vehicle (UAV). A drone fitted with downward facing sensors, is used to capture data while flying above the ground. The sensors may be RGB (red, green and blue) or multispectral cameras. Many drones use LIDAR payloads (a remote sensing method) for collecting data. In drone survey overlapping photographs are taken to generate 3 D models of earth surface and to generate panoramic views.

Drone surveying is used for 2-D and 3-D mapping, land surveys, topographic surveys and photogrammetry. The drones can capture the data at vantage points which are not accessible to humans, especially challenging terrains. Human operators are not required to measure points in hazardous locations physically.

UNIT SUMMARY

- **Remote Sensing**: Remote sensing is a technique of obtaining information about a target by collecting data with the help a device that is not in contact with the target.
- **Passive remote sensing**: It takes the electromagnetic radiations from the sun and measure the reflected/emitted radiations from the target.
- **Active remote sensing**: It takes the electromagnetic radiations from the its own source and measure the reflected/emitted radiations from the target.
- **The Geographical Information System (GIS):** GIS is a system which analyses geographical information obtained by remote sensing, through computer hardware and software.
- **Components of GIS:** GIS consists of following components
  (i) Hardware
  (ii) Software
  (iii) Man power
  (iv) Information
- **The Global Positional System (GPS):** Deals with the techniques which determine global position of an object or a feature using data obtained through remote sensing.
- **Principle of GPS:** The satellites send the position and time information to the receiver at fixed intervals. The signals are sent to the receiver by radio waves. The time difference between the signal sent from satellite to the signal received by the receiver determines the distance between the receiver and the satellite. The receiver locates its position by obtaining the signals from at least three satellites.
- **Elements of GPS:** GPS consists of following parts

(i) Satellites (ii) Ground Stations (iii) Receivers

- Application of GIS

  The GIS has applications in folloeing fields.

  (i) Agriculture
  (ii) Urban and Town Planning
  (iii) Oil Spill
  (iv) Disaster Management
  (v) Navigation and Mapping.
  (vi) Dam and Reservoir Site location
  (vii) Vegetation and Deforestation Management
  (viii) Business and Marketing Strategies

EXERCISES

Multiple Choice Questions.

1. In the Remote sensing which kind of waves are used?
   (a) X – Ray waves
   (b) Solar waves
   (c) Gamma- rays
   (d) Electro-magnetic waves
2. Which of the following isn't a principle of remote sensing?
   (a) Interaction of energy with satellite
   (b) Interaction of energy with atmosphere
   (c) EM spectrum
   (d) EM energy
3. Following is not the element of GPS.
   (a) Satellite
   (b) Ground station
   (c) Receiver
   (d) Reflector
4. The relation between wavelength, velocity and frequency is __________
   (a) $\lambda = \frac{c}{f}$
   (b) $\lambda = \frac{c}{h}$
   (c) $\lambda = h\frac{c}{f}$
   (d) $\lambda = \frac{c}{r}$
5. What is the full form of GIS
   (a) Global Information System
   (b) Geographical Information System
   (c) Geological Information System
   (d) Geo-synthetic Information System
6. What is the full form of GPS
   (a) Global Positional System
   (b) Geographical Positional System
   (c) Geological Positional System
   (d) Geo-synthetic Positional System
7. Following is not the component of GIS
   (a) Hardware
   (b) Software
   (c) Information
   (d) User

8. which of the following is an example of hardware component in GIS.

(a) CAD (b) Arc GIS
(c) Keyboard (d) Digitalization

9. In passive remote sensing, the electromagnetic radiation are taken from

(a) Sun (b) own source
(c) Nuclear energy (d) Electric energy

10. Which of the following isn't related to GIS software's?

(a) ARC/INFO (b) PAMAP
(c) STAAD Pro (d) SPANS

11. Which of the following is having same principle of determining the position in GPS?

(a) Compass (b) Resection
(c) Trisection (d) Traversing

12. Satellite generates which type of signals?

(a) Visible rays (b) Radio waves
(c) Cosmic waves (d) X-rays

13. Which of the following is used to locate the coordinates of an object?

(a) RS (b) GIS
(c) GPS (d) IRS

14. In the active remote sensing, the electromagnetic radiations are taken from its own source.

(a) False (b) True

15. The visible wavelength covers range from ....

(a) 0.4 to 0.7 centi- meter (b) 0.4 to 0.7 micrometer
(c) 0.4 to 0.7 nanometer (d) 0.4 to 0.7 meter

ANSWERS															
1.	(d)	2.	(a)	3.	(d)	4.	(a)	5.	(b)	6.	(a)	7.	(d)	8.	(c)
9.	(a)	10.	(c)	11.	(b)	12.	(b)	13.	(c)	14.	(b)	15.	(b)		

Short and Long Answer Type Questions Short Questions:

1. Explain remote sensing in brief.
2. What are the components of remote sensing system?
3. Why remote sensing is used in land use/ land cover mapping?
4. How are damages reduced by remote sensing in disaster management?
5. What are the components of GPS?
6. Explain principle of GPS.
7. What is GIS?

8. What are the components of GIS?
9. List the uses of GIS
10. List the GIS software
11. Explain drone surveying in brief.

Long Questions

1. Describe the various components of remote sensing
2. Discuss the applications of remote sensing in detail.
3. What is GIS? Describe its components and applications
4. Write notes on following
   (i) GPS (ii) GIS software (iii) Drone surveying

PRACTICAL

Experiment No. – 1

Objective: Use GPS to locate the coordinates of a station.

Required Equipment:

GPS receiver

Procedure:

Start the GPS.

Click on the "setup" icon and go in "system" and click on satellite system

Choose the language and set the unit like distance in meter etc.

Click on position format and choose degree, minutes and seconds.

Click on markway point option and press "OK". After pressing on "OK", position will be on

Give the name of the point and click on "Done".

Now the GR (Grid Reference) is Saved, Coordinates will be displayed by clicking on GR option.

Results:

The coordinate of desired point is............................

KNOW MORE

- **Law of radiation and their relevance in remote sensing** – https://youtu.be/soahAd_GmnY
- **Remote sensing image representation - https://youtu.be/-jl20ok55sc**
- **Various Remote sensing platform** - https://youtu.be/Yt54A2KIXp8
  1. Airborne Platform
  2. Space borne Platform
  3. Satellite
- **Remote Sensing Sensors** – It is mainly two types

1. **Passive sensors** – It has two types
    (i) Non Scanning
    (ii) Scanning
2. **Active Sensors** - It has two types
    (i) Non Scanning
    (ii) Scanning

- **Sensors Parameters**
    (i) Spatial Resolution
    (ii) Spectral Resolution
    (iii) Radiometric Resolution
    (iv) Temporal Resolution
- **Image Processing**

    Digital image processing may involve various procedure including correcting and formatting of the data, digital enhancement to the better visual interpretation.
- **Data structure** - In GIS, the data is stored in following structures
    (i) Relational data structure
    (ii) Hierarchical data structure
    (iii) Network data structure
- **Errors in GIS** – GIS Error can be classified into following groups.
    1. **Error associates with data**
        (i) Age of data
        (ii) Map Scale
        (iii) Density of observations
        (iv) Data Inaccuracy
        (v) Inaccuracy
    2. **Errors associated with processing**
        (i) Map digitization errors
        (ii) Rasterisation errors
        (iii) Spatial integration errors
        (iv) Generalization errors
        (v) Attribute mismatch errors
        (vi) Misuse of logic
- **Spectral Reflectance of Vegetation, Soil and Water**
    1. https://www.scribd.com/presentation/438578619/Spectral-Reflectance-of-Vegetation-Soil-and-Water
    2. https://r.search.yahoo.com/_ylt=Awr1UbpDUHZjTDAONMO7HAx.;_ylu=Y29sbwNzZzMEcG9zAzEEdnRpZAMEc2VjA3Ny/RV=2/RE=1668726979/RO=10/RU=https%3a%2f%2fwww.coursehero.com%2ffile%2f54018624%2f2-Spectral-reflectancepdf%2f/RK=2/RS=sgSCBWPpvIc7J3i3oYB4up1BLV8-

- Bhuvan (NRSC) Website for browsing the data sets
  https://r.search.yahoo.com/_ylt=AwrPqntuUXZj9DwPgiq7HAx.;_ylu=Y29sbwNzZzMEcG9zAzEEdnRpZAMEc2VjA3Ny/RV=2/RE=1668727278/RO=10/RU=https%3a%2f%2fbhuvan.nrsc.gov.in%2f/RK=2/RS=fbYCgShvU2pc2lRTHJ3ffSGmAVQ-
- **Navigation Systems**
  https://www.google.co.in/url?sa=t&rct=j&q=&esrc=s&source=web&cd=&ved=2ahUKEwi_rKbixbX7AhUQdXAKHaDPAHQQFnoECAgQAw&url=https%3A%2F%2Fwww.gps.gov%2Fsystems%2Fgnss%2F&usg=AOvVaw00jIj_Hmm8l7KsKuLkQkAK&cshid=1668699523482020

REFERENCES AND SUGGESTED READINGS

1. Basudeb Bhatta, Remote Sensing and GIS, Oxford Higher Education, 2021.
2. Shivam Pandey, Dr. Shashikant Tripathi, Basic concept of Remote Sensing, GPS and GIS, Sankalp Publication, 2020.
3. George Joseph, C Jegananathan, Fundamentals of Remote Sensing, The orient Blackswan, 2018.
4. B.C. Punmia, Surveying – III Higher Surveying, Laxmi Publications (P) Ltd, 2017.
5. Gopi A, Satikumar R., Advance surveying, Pearson Education India, 2017.
6. Lillesand T.M. and Kiefer R.W, Remote Sensing and image interpretation, Wiley Publication, 2015
7. Remote Sensing Essentials –
   https://youtube.com/playlist?list=PLLy_2iUCG87CDlroZBlwwBllYwz7KxVtA
8. Introduction to Remote Sensing and GIS
   https://youtu.be/tqpHVsUfg6o

Dynamic QR Code for Further Reading

Following QR codes are given for further study of plan Remote sensing, GPS and GIS

Remote Sensing and GIS

Basic of GPS, Receivers, Principles and Application

GPS Surveying and Applications

GPS System

Remote Sensing and GIS

Satellite Remote sensing

REFERENCES FOR FURTHER LEARNING

1. Dr. K. R. Arora, Surveying Vol. I, Standard Book House, 2019.
2. A. M. Chandra, Plane Surveying, New Age International Publishers, 2015.
3. Subramanian, Surveying and Levelling , Oxford University Press, 2012.
4. C. Venkatramaiah, Textbook of Surveying, University Press, 2011
5. John Uren & Bill Price, Surveying for Engineers, Palgrave Macmillan, 2010.
6. S. K. Roy, Surveying Fundamentals and Practices Prentice Hall India Learning Private Limited, 2010.
7. M. James, Anderson Edward Mikhail, Surveying Theory and Practic, Tata McGraw Hill, 2000.
8. Arthur R Benton, Philip J Taety, Elements of Plane Surveying, McGraw-Hill Education, 2000.
9. James M. Anderson, Surveying Theory and Practice,McGraw Hill, 1998.
10. IRC: 38 – Design Tables for Horizontal Curves for Highways
11. IRC SP: 23 – Vertical Curves for Highways

e-Resources:

1. http://www.bis.org.in/sf/wrd/p_449.pdf
2. http://www.bis.org.in/sf/wrd/WRD10(491).pdf
3. http://www.bis.org.in/sf/wrd/WRD10(491).pdf
4. http://sbq.com.au/member/board-publications/code-of-practice/
5. http://usa.autodesk.com/adsk/servlet/pc/index?id=3091031&siteID=123112
6. http://www.cadacademynoida.com/?page=civileng3
7. http://www.sitetopo.com
8. https://www.slideshare.net/silashah12/gps-and-remote-sensing
9. https://medium.com/@srinivastanniru/gis-remote-sensing-and-gps-what-is-the-difference-29d7ee71ccf6
10. https://nptel.ac.in/courses/105103193

REFERENCES FOR FURTHER READING

CO AND PO ATTAINMENT TABLE

Course outcomes (COs) for this course can be mapped with the programme outcomes (POs) after the completion of the course and a correlation can be made for the attainment of POs to analyze the gap. After proper analysis of the gap in the attainment of POs necessary measures can be taken to overcome the gaps.

Table for CO and PO attainment

Course Outcomes	Attainment of Programme Outcomes (1- Weak Correlation; 2- Medium correlation; 3- Strong Correlation)						
	PO-1	PO-2	PO-3	PO-4	PO-5	PO-6	PO-7
CO-1							
CO-2							
CO-3							
CO-4							
CO-5							
CO-6							

The data filled in the above table can be used for gap analysis.

INDEX